GREEN FACILITIES

McGRAW-HILL'S GREENSOURCE SERIES

Attmann
Green Architecture: Advanced Technologies and Materials

Gevorkian
Alternative Energy Systems in Building Design
Large-Scale Solar Power System Design: An Engineering Guide for Grid-Connected Solar Power Generation
Solar Power in Building Design: The Engineer's Complete Design Resource

GreenSource: The Magazine of Sustainable Design
Emerald Architecture: Case Studies in Green Building

Haselbach
The Engineering Guide to LEED–New Construction: Sustainable Construction for Engineers, 2d ed.

Luckett
Green Roof Construction and Maintenance

Melaver and Mueller (eds.)
The Green Building Bottom Line: The Real Cost of Sustainable Building

Nichols and Laros
Inside the Civano Project: A Case Study of Large-Scale Sustainable Neighborhood Development

Winkler
Green Facilities: Industrial and Commercial LEED Certification
Recycling Construction & Demolition Waste: A LEED-Based Toolkit

Yellamraju
LEED–New Construction Project Management

Yudelson
Green Building Through Integrated Design
Greening Existing Buildings

GREEN FACILITIES

INDUSTRIAL AND COMMERCIAL LEED CERTIFICATION

GREG WINKLER, AIA

New York   Chicago   San Francisco   Lisbon   London   Madrid
Mexico City   Milan   New Delhi   San Juan   Seoul
Singapore   Sydney   Toronto

The McGraw·Hill Companies

Cataloging-in-Publication Data is on file with the Library of Congress

Copyright © 2011 by The McGraw-Hill Companies, Inc. All rights reserved. Printed in the United States of America. Except as permitted under the United States Copyright Act of 1976, no part of this publication may be reproduced or distributed in any form or by any means, or stored in a data base or retrieval system, without the prior written permission of the publisher.

1 2 3 4 5 6 7 8 9 0   DOC/DOC   1 7 6 5 4 3 2 1

ISBN 978-0-07-174453-9
MHID 0-07-174453-3

Sponsoring Editor	**Project Manager**
Joy Evangeline Bramble	Anupriya Tyagi, Glyph International
Editing Supervisor	**Copy Editor**
Stephen M. Smith	Nidhi Chopra, Glyph International
Production Supervisor	**Art Director, Cover**
Pamela A. Pelton	Jeff Weeks
Acquisitions Coordinator	**Composition**
Alexis Richard	Glyph International

Printed and bound by RR Donnelley/Crawfordsville.

McGraw-Hill books are available at special quantity discounts to use as premiums and sales promotions, or for use in corporate training programs. To contact a representative, please e-mail us at bulksales@mcgraw-hill.com.

 The pages within this book were printed on acid-free paper containing 100% postconsumer fiber.

To my son, Tristan.
Sometimes happiness slips in through a door
you didn't know you had left open.

About *GreenSource*

A mainstay in the green building market since 2006, *GreenSource* magazine and GreenSourceMag.com are produced by the editors of McGraw-Hill Construction, in partnership with editors at BuildingGreen, Inc., with support from the United States Green Building Council. *GreenSource* has received numerous awards, including American Business Media's 2008 Neal Award for Best Website and 2007 Neal Award for Best Start-up Publication, and FOLIO magazine's 2007 Ozzie Awards for "Best Design, New Magazine" and "Best Overall Design." Recognized for responding to the needs and demands of the profession, *GreenSource* is a leader in covering noteworthy trends in sustainable design and best practice case studies. Its award-winning content will continue to benefit key specifiers and buyers in the green design and construction industry through the books in the *GreenSource* Series.

About McGraw-Hill Construction

McGraw-Hill Construction, part of The McGraw-Hill Companies (NYSE: MHP), connects people, projects, and products across the design and construction industry. Backed by the power of Dodge, Sweets, *Engineering News-Record* (*ENR*), *Architectural Record*, *GreenSource*, *Constructor*, and regional publications, the company provides information, intelligence, tools, applications, and resources to help customers grow their businesses. McGraw-Hill Construction serves more than 1,000,000 customers within the $4.6 trillion global construction community. For more information, visit www.construction.com.

About the International Code Council

The International Code Council (ICC), a membership association dedicated to building safety, fire prevention, and energy efficiency, develops the codes and standards used to construct residential and commercial buildings, including homes and schools. The mission of ICC is to provide the highest quality codes, standards, products, and services for all concerned with the safety and performance of the built environment. Most United States cities, counties, and states choose the International Codes, building safety codes developed by the International Code Council. The International Codes also serve as the basis for construction of federal properties around the world, and as a reference for many nations outside the United States. The Code Council is also dedicated to innovation and sustainability, and a Code Council subsidiary, ICC Evaluation Service, issues Evaluation Reports for innovative products and reports of Sustainable Attributes Verification and Evaluation (SAVE).

Headquarters: 500 New Jersey Avenue NW, 6th Floor, Washington, DC 20001-2070

District Offices: Birmingham, AL; Chicago, IL; Los Angeles, CA

1-888-422-7233; www.iccsafe.org

About the Author

Greg Winkler, AIA, is the executive director of a regional construction trade organization. An architect and project manager with more than 25 years of experience in affordable housing, office buildings, industrial/commercial, and retail construction, he is the co-author of *Construction Administration for Architects* and author of *Recycling Construction & Demolition Waste: A LEED-Based Toolkit.*

CONTENTS

PREFACE

Green sells. More than any other factor, the business community has embraced sustainability because it is marketable. Consumers have long accepted the idea of recycling their household waste, and have even shown an increased willingness to purchase products with recycled content—as long as they are competitive in cost with products made from virgin material. What is new is that customers are now interested in supporting businesses that not only use recycled content in their products, but also utilize sustainable principles in the manufacture of those products and the overall operation of their businesses.

In the rush to market themselves or their products as green, some corporations have adopted the unfortunate practice of "greenwashing," or making sustainability claims that either cannot be documented or have no basis in fact. That even some multinational corporations with strong reputations would make such claims is unfortunate, and devalues the true and honest commitment to sustainability practiced by so many other businesses, both large and small. Greenwashing exists, of course, because even the false claim of green yields profit. The shame is that true greenness, the honest practice of sustainability, can yield even greater profits and productivity.

That is where the simplicity of sustainability ends, because sustainability is anything but simple. Energy-efficient operations vary dramatically with the type and nature of the business. The potential savings available among manufacturing, professional, or retail businesses is quite different, and must be weighed against comfort, safety, and productivity. Sustainability is also a moving target, and like so many other aspects of running a business, requires constant assessment and adjustment. Energy-efficiency standards are always increasing, and growth in many locales will tax the water, clean air, and wastewater capacities of those communities. No aspect of sustainability stays constant for very long. But then, what aspect of the operation of a business is ever constant?

For all its vexing aspects, running green is little different from running lean. Once implemented, it is a way of life, a way of operating smarter and with less waste. The businesses that embrace this simple fact will be more profitable and more competitive. They will have embraced the heart of sustainability.

Greg Winkler, AIA

ACKNOWLEDGMENTS

I would like to express my appreciation to Joy Evangeline Bramble, Senior Editor, McGraw-Hill Professional, for her continued interest in my writing. I also appreciate the professionalism of the entire McGraw-Hill Professional team, particularly Pamela Pelton and Stephen Smith.

My thanks go to the International Code Council, Inc., for its co-branding of this book and its strong commitment to improving the built environment around the world.

I would like to acknowledge Hamid Naderi, PE, Deputy Senior Vice President of Business & Product Development for the International Code Council, for his review of and comments on the manuscript for this book.

I also appreciate the copy editing and production talents of Anupriya Tyagi and the staff at Glyph International.

GREEN FACILITIES

GREEN FACILITIES

Green facilities are smart facilities. They are businesses that control their costs through focused attention on reduced energy consumption, enhanced equipment efficiency, consistent maintenance, and more flexible building and human resource management. In the sense that business environmental sustainability is largely measured in resource efficiency, businesses have been practicing sustainability for a long time under the name of *cost reduction*. A business that did not routinely look for ways to produce their products or services less expensively was destined to be overtaken by producers who operated more efficiently and sold their wares for less. This aspect of green practices is not new, though the tools available to today's managers for assessing and implementing cost reduction measures are vastly greater than those of even a decade ago. What is new, and what this book addresses, is the beginning of a new era of looking at a wider range of sustainability factors—including facilities, human resources, equipment, and operations—in a comprehensive manner as part of an overall sustainability program. The new reality of sustainable management means assessing the effects of facility and equipment changes on employee productivity, production efficiency, energy consumption, and a host of other interrelated factors:

- How does implementing an employee ride-share program affect employee working hours, productivity, and building operation costs?
- Can the use of compressed air be reduced without losing productivity, and is it worth the cost?
- Does a redesign of the steam supply system for greater efficiency allow for cost-effective expansion when orders increase?
- How does allowing employees to open windows during temperate periods affect absenteeism and liability associated with asthma sufferers?

Managing in the green era requires broader vision, more thoughtful analysis, and a healthy dose of prognostication. Sustainable management is part analytical and part intuitive, a blend of business benefits and greater good. Most of all, sustainability is now an established movement with public and employee support. Creating greener facilities is a necessity for managers, and they would do well to follow the advice of Francis Bacon: "Things alter for the worse spontaneously, if they be not altered for the better designedly."

What Is Sustainability?

The most common definition of sustainability is that adopted by the Brundtland Commission of the United Nations in March 1987: "Sustainable development is development that meets the needs of the present without compromising the ability of future generations to meet their own needs." This very broad definition is helpful to governments and the public, but provides little guidance to business or facility managers. The Council of Smaller Enterprises (COSE) offers a crisper definition in their mission statement: "Sustaining and supporting the small business community through improved economic, environmental, and community design strategies leading to increased implementation of energy conservation, recycling, and employee wellness programs."[1]

The Benefits of Sustainability

The greening of global business has, in some ways, occurred with surprising speed. Long-time backers of environmental movements would certainly disagree, citing the origins of their movement in the 1960s and the long, difficult road to persuading American consumers and businesses that recycled products have high quality and marketability. What has occurred in the last decade, however, has been a widespread embracing of sustainability benefits by global consumers. While the American public is the newcomer to this party (European consumers having embraced green concepts long ago), they have done so with gusto. A 2008 Gallup Poll revealed that 72 percent of Americans avoided using products that harm the environment, with 86 percent of Americans stating that they are currently recycling household waste such as newsprint and glass.[2] Sustainability may no longer represent a marketing advantage so much as a consumer expectation.

Companies that have embarked on sustainability audits, either limited or comprehensive, report improvements both in public perception and in their bottom line. While the need to at least appear green has prompted a fair amount of "greenwashing" (false or unproven sustainability claims), responsible corporations have embraced sustainability as a way to improve their overall operations and facility competitiveness. Among them:

■ Gundersen Lutheran Hospital of La Crosse, Wisconsin, conducted an audit of their facilities to look for improvements that could be implemented quickly, said Jerry Arndt, Senior Vice President of Business Services. "The most responsible thing you can do is reduce the amount of energy you need," Arndt said. "So we looked in-house for improvements before we looked at renewables." Among other items, Gunderson found their facility included 300 exhaust fans that were running full time and could be shut off for 12 hours each day. Another part of their audit determined that 60 percent of the hospital's energy expenses were incurred for steam production. As a result, the hospital began a program of replacing and repairing steam traps to improve efficiency.[3]

■ Proctor and Gamble's energy-efficiency program reduced water consumption by 52 percent, energy usage by 48 percent, CO_2 emissions by 52 percent, and waste disposal by 53 percent in their production facilities.[4]

■ In 2004, Timberland began energy-efficiency improvements that resulted in savings of over 40 percent in energy and emissions at the company's largest facilities. In 2006, they began using solar photovoltaic (PV) arrays at their Ontario, California, Distribution Center. That facility is now 60 percent powered by their PV system.[5]

■ Walmart has committed to eliminating landfill waste at their stores and Sam's Club facilities by 2025. Between February 2008 and January 2009, they reduced landfill waste from their stores by more than 57 percent. The company has established three clear, ambitious green goals[6]:

1 Being supplied 100 percent by renewable energy,
2 Creating zero waste, and
3 Selling products that sustain people and resources.

All of these companies, in ways ranging from limited to expansive, have employed green practices to benefit their image and competitiveness. Panelists at the 2009 West End Business Improvement Association Sustainability Expo in Vancouver, BC, agreed on five key factors that are driving businesses to become more sustainable[7]:

■ Consumer demand
■ Competition and branding
■ Cost savings
■ Current employee morale and new employee demands
■ Regulations

Sustainability simply makes good business sense. Managing facilities in a sustainable manner means that, especially in times of economic uncertainty, a company is working to reduce costs and achieve operating efficiencies. Along with whatever marketing or public image benefits it derives from green practices, businesses also achieve competitive advantage and sustained growth from the benefits of using less energy and producing goods more efficiently. See Fig. 1.1 for a summary of potential benefits of adopting sustainable practices in commercial facilities.

This is a reality already recognized by a high percentage of companies. A survey of U.S. companies showed an increasing trend toward adopting sustainable principles in workplace management[8]:

■ 58 percent of U.S. employers surveyed have formal green workplace programs
■ 85 percent use webinars or teleconferencing
■ 78 percent have green communication programs to reduce paper usage
■ 72 percent use online human resources communications

- Reduced material costs
- Reduced waste costs
- Reduced energy costs
- Marketing advantages
- Product branding benefits
- Better employee morale
- Enhanced competitiveness

Figure 1.1   Potential benefits of sustainability.

- 58 percent have internal communication programs that offer employees green tips
- 57 percent use online sustainability plan summaries
- 57 percent offer telecommuting
- 52 percent offer a ride-share program

Sustainability in commercial and industrial facilities begins with efficiency. Efficiency is an inexpensive form of renewable energy, generating a higher return on investment than any other improvement. Reducing overall utility usage and waste, diverting waste from landfills to recyclers, utilizing more electronic communication in lieu of paper, and providing healthier work environments for employees are a few of the most common methods that companies employ to improve efficiency and reduce costs. Efficiency is also an ideal way to include employees and consultants in the process of achieving energy and cost savings. Many companies have found their employees will enthusiastically participate in green reward programs that offer recognition and a small financial inducement for identifying workplace economy measures.

The ICC and Sustainability

The International Code Council (ICC) is the preeminent code authority in the country, dedicated to creating model codes that promote building safety, fire prevention, and energy efficiency for residential and commercial buildings. ICC codes are in use, to varying extents, in all fifty states and are used as the basis for the codes of several other countries. The ICC, in partnership with the American Institute of Architects (AIA), the U.S. Green Building Council (USGBC), and three other organizations, has developed the International Green Construction Code (IgCC) slated for release in early 2012. This initiative includes collaboration among a wide range of professional, trade, and green building leaders, as well as outreach and feedback from the general public. The resulting code will cover all aspects of sustainability in the built environment, including drawing from existing codes and standards, to create a single universal code. The IgCC will apply to new construction and renovations, will link in closely with existing American Society of Testing & Materials (ASTM International) standards, and will provide a regulatory framework to assist municipalities and code officials in understanding and administering green construction.

GOALS OF THE IgCC

The goal of the IgCC is to decrease energy usage and carbon footprints along with addressing several other issues:

- The code addresses site development and land use, including preservation of natural and material resources as part of the process.
- Enforcement of the code will improve indoor air quality and support the use of energy-efficient appliances, renewable energy systems, water resource conservation, rainwater collection and distribution systems, and the recovery of used water (also known as gray water).
- The IgCC emphasizes building performance, including requirements for building system performance verification and building owner education to ensure that the best energy-efficient practices are being carried out.
- A key feature of the new code is a section devoted to "jurisdictional electives," which will allow customization of the code beyond its baseline provisions to address local priorities and conditions.

Additional information regarding the IgCC is available at http://www.iccsafe.org/. See Table 1.1 for a list of IgCC sustainable code categories.

SUSTAINABLE ATTRIBUTES VERIFICATION AND EVALUATION™ (SAVE™) PROGRAM

The ICC subsidiary, the ICC Evaluation Service (ICC-ES), has created the ICC-ES SAVE™ Program to verify manufacturers' claims regarding the sustainable characteristics of their products. One of the biggest problems arising out of the popularity of sustainable construction is the practice of greenwashing, or making environmental sustainability claims for a product that cannot be documented or are patently false. The purpose of the SAVE™ program is to give manufacturers the opportunity to voluntarily document the sustainable attributes of their products. Once approved by ICC-ES, this

TABLE 1.1 IgCC KEY CATEGORIES

- Natural resource conservation and responsible land use and development
- Material resource conservation and efficiency
- Energy conservation, efficiency, and earth atmospheric quality
- Water resource conservation and efficiency
- Indoor Environmental Quality (IEQ) and comfort
- Building operation, maintenance, and owner education
- Existing buildings

TABLE 1.2   SAVE EVALUATION GUIDELINES

EG101—Evaluation guideline for determination of recycled content of materials

EG102—Evaluation guideline for determination of biobased material content

EG103—Evaluation guideline for determination of solar reflectance, thermal emittance, and solar reflective index of roof covering materials

EG104—Evaluation guideline for determination of regionally extracted, harvested, or manufactured materials or products

EG105—Evaluation guideline for determination of volatile organic compound (VOC) content and emissions of adhesives and sealants

EG106—Evaluation guideline for determination of VOC content and emissions of paints and coatings

EG107—Evaluation guideline for determination of VOC content and emissions of floor covering products

EG108—Evaluation guideline for determination of formaldehyde emissions of composite wood and engineered wood products

EG109—Evaluation guideline for determination of certified wood and certified wood content in products

documentation can help those seeking to qualify for points under major green rating systems, LEED or Green Globes program. See Table 1.2 for a list of SAVE program evaluation guidelines.

LEED Certification

Leadership in Energy and Environmental Design (LEED)® is a third-party certification program and a widely accepted benchmark for the design, construction, and operation of sustainable buildings. Developed by the USGBC in 1998 through a committee process involving a wide range of nonprofit, industry, and governmental groups, LEED serves as a design and construction template for sustainable buildings of all types and sizes. As of 2009, USGBC estimated that more than 4.5 billion square feet (4.18 billion square meters) of building area has been designed under the LEED program.

The certification process, administered by a USGBC spin-off organization called the Green Building Certification Institute (GBCI), determines the appropriateness of buildings for LEED certification through an online submission process, using templates for each of the seven types of construction recognized under the LEED system. LEED certification is available for the following building types: general new construction, major renovation, existing buildings, commercial interiors, core and shell, schools, and homes. As of this writing, LEED certifications for retail space, neighborhood development, and healthcare facilities are currently in development.

LEED NEW CONSTRUCTION

LEED New Construction (LEED NC) is the category most often used by businesses for LEED certification. This category covers, as the name implies, almost all new construction work (see the core and shell exception below), and major renovation work as well. According to USGBC, "A major renovation involves major HVAC renovation, significant envelope modifications, and major interior rehabilitation." In cases with lesser degrees of renovation not meeting the criteria for LEED NC, owners and contractors should use the LEED for Existing Buildings: Operation and Maintenance criteria (LEED EB).

To qualify for certification, projects must meet certain prerequisites and earn additional performance points in six base categories of sustainable design. The six categories are:

- Sustainable Sites (SS)
- Water Efficiency (WE)
- Energy and Atmosphere (EA)
- Materials and Resources (MR)
- Indoor Environmental Quality (IEQ)
- Regional Priority (RP)

An additional category, Innovation in Design (ID), addresses unusual or exceptional situations, as well as design situations not covered under the six environmental categories. See Table 1.3 for a summary of the points and prerequisites associated with each category.

TABLE 1.3   LEED NC CATEGORIES

- Sustainable Sites: 26 possible points
  - Prerequisite 1: Construction activity pollution prevention

- Water Efficiency: 10 possible points
  - Prerequisite 1: Water use reduction

- Energy and Atmosphere: 35 possible points
  - Prerequisite 1: Fundamental commissioning of building energy systems
  - Prerequisite 2: Minimum energy performance
  - Prerequisite 3: Fundamental refrigerant management

- Materials and Resources: 14 possible points
  - Prerequisite 1: Storage and collection of recyclables

- Indoor Environmental Quality: 15 possible points
  - Prerequisite 1: Minimum indoor air quality performance
  - Prerequisite 2: Environmental Tobacco Smoke (ETS) control

- Innovation in Design: 6 possible points

- Regional Priority: 4 possible points

LEED projects can earn up to 100 base points, plus an additional six Innovation and Design Process credits and up to four Regional Priority points for obtaining products within a 500-mile (805-kilometer) radius of the construction site. Projects are recognized in four progressive categories, according to the number of points they earn:

- Certified Project: 40–49 points
- Silver Project: 50–59 points
- Gold Project: 60–79 points
- Platinum Project: 80 points and above

The USGBC New Construction (NC) standards may apply to new buildings or to renovation projects where a significant portion of the facility is undergoing renovation. The LEED program does make a distinction, however, regarding projects that are designed and constructed to be partially occupied by the owner or developer, with the rest being occupied by tenants. In these projects, USGBC claims that the owner or developer has direct influence over the self-occupied portion of the building. To pursue certification under the LEED NC category, the owner must therefore occupy 50 percent or more of the building's leasable area. Where this is not the case, the project is not suitable for LEED NC credits and the owner should pursue LEED for Core & Shell certification.

Total LEED program-related costs vary with the project type and size. Registration of a project with GBCI costs approximately $450 (for USGBC members) and $600 (for non-members). LEED certification costs vary with project size, but the USGBC states that the average cost is around $2,000. Indirect costs may include independent LEED consultants and building commissioning. These costs vary dramatically based on the size and complexity of the project.

DOCUMENTATION REQUIREMENTS

LEED certification requires extensive documentation to prove that the project was completed in accordance with the requirements. LEED provides customized online templates (fillable forms) for LEED-accredited professionals to use for documenting each prerequisite and credit. Additional documentation is sometimes required beyond the information presented on the form, and the level of documentation can be rather extensive. The documentation for recycled content and regional materials, for instance, can require obtaining information from numerous outside suppliers for projects with multiple components from different sources.

Because the contractor is responsible for construction waste management, businesses will need to rely on him for documenting compliance with this prerequisite. This will require the contractor to tabulate the total waste material, quantities diverted from landfills, and the means through which they were diverted.

A portion of the credits in each application will be audited, and the company should be prepared with backup documentation for any credit claimed, whether required as part of the initial application or not. Certification is now administered by the GBCI through a network of professional, third-party certification bodies. To register a project for LEED certification, visit www.gbci.org.

LEED FOR EXISTING BUILDINGS

The LEED EB rating system states that it "helps building owners and managers measure operations, improvements, and maintenance on a consistent scale, with the goal of maximizing operational efficiency while minimizing environmental impacts." LEED EB includes criteria for whole-building cleaning and maintenance activities, including recycling programs, chemical use, exterior maintenance programs, and building systems improvements. The LEED EB criteria are intended to apply both to existing buildings seeking LEED certification for the first time as well as to projects previously certified under LEED for New Construction, Schools, or Core and Shell criteria. The LEED EB category requires that a number of requirements be met prior to application for certification. See Table 1.4 for a listing of the Existing Building category pre-application requirements.

The heart of the LEED EB criteria is an operations and maintenance (O&M) guide provided as a managerial tool to assist businesses in monitoring the ongoing operations and maintenance of existing commercial and institutional buildings. LEED EB uses this guide as a tool to identify and reward building O&M best practices. It is also intended as a useful template for businesses in identifying opportunities to reduce energy and utility consumption, improve their indoor environment, and identify and correct operating inefficiencies. The rating system is targeted at single buildings and requires three months of operational data prior to the initial application. Buildings must be in operation for twelve months prior to certification. The O&M template is intended to address whole-building performance as opposed to that of individual tenant spaces. Any building construction or renovation must be complete at least three months before LEED EB certification can be requested. See Table 1.5 for a list of LEED EB certification major compliance categories.

GBCI offers the following tips for managers seeking LEED certification:

1 Review the LEED rating system to assess how many credits are possible.
2 Choose a target LEED certification level: Certified, Silver, Gold, or Platinum.

TABLE 1.4   LEED EXISTING BUILDING CERTIFICATION PREAPPLICATION REQUIREMENTS

- The building must be at least 75% occupied for 12 months preceding application.
- For residential facilities such as hotels, apartments, or condominiums, average occupancy over the preceding 12 months cannot be less than 75% of the total floor area.
- The project scope must include 100% of the total floor area.
- The building must comply with all local, state, and federal environmental laws.

TABLE 1.5   LEED EXISTING BUILDINGS MAJOR CATEGORY SUMMARY

Refer to *www.usgbc.org* to purchase the full document

Sustainable Sites (9 points)
- Exterior and hardscape management
- Pest, landscape, and erosion management
- Protect or restore open space
- Alternative transportation
- Roof and nonroof heat island reduction
- Light pollution reduction

Water Efficiency (4–10 possible points)
- Water performance measurement
- Fixture and fitting efficiency
- Water efficient landscaping
- Cooling tower management

Energy & Atmosphere (13–30 possible points)
- Energy efficiency performance
- Refrigeration management
- Building commissioning
- Performance measurement
- Renewable energy
- Emissions reduction

Materials & Resources (9–14 possible points)
- Solid waste management
- Purchasing and ongoing consumables
- Durable goods
- Facility renovations and alterations

Indoor Environmental Quality (16–20 possible points)
- Air intake and exhaust systems
- Tobacco smoke control
- Green cleaning practices
- Indoor air quality best management practices
- Occupant comfort factors (lighting, thermal, and views)
- Pest management practices

Innovation in Operations (4–7 possible points)
- Innovative operations
- LEED accredited professional
- Documenting sustainable building cost impacts

3 Assess what equipment will require upgrades.

4 Assign company personnel responsibility for achieving the credits and for preparing green policies.

5 Prepare a budget.

6 Prepare a schedule and project management plan.

7 Register the project through GBCI and use resources available online.

LEED LIMITATIONS

As businesses grow more sophisticated regarding sustainable practices, there are questions as to whether LEED is the best vehicle for facility managers to use in measuring their progress or seeking certification for their achievements. Architect and sustainable design educator Warren Wagner has summarized LEED as "minimum requirements at minimum investment." Many buildings can qualify for LEED in ways that are relatively cheap and easy, neglecting the overall goal of significantly reducing the building's impact. Wagner also does not believe that the LEED program is ambitious enough to push the industry toward meeting the ambitious 2030 Challenge established by the AIA and other international organizations of architects in 2006. This initiative calls for all new buildings and major renovations to reduce greenhouse gas emissions by 50 percent by 2010 and become carbon neutral by 2030.

LEED also does not address high-performance issues of durability, weather, or fire resistance. As a result, a LEED-compliant building may not be as economical or sustainable in terms of life-cycle costs as buildings that do not meet the LEED requirements.[9]

Other Sustainability Programs

Although the USGBCs LEED program is the most well known, other programs have been developed by national organizations to promote sustainable construction. These programs, some of which include certification components, incorporate aspects of recycling construction and demolition waste to varying extents.

GREEN GLOBES

The Green Globes program is a voluntary certification program administered by the Green Building Initiative, a broad-based consortium of industry, government, and nonprofit representatives who modified an early Canadian program into an online resource that is promoted as a more streamlined and interactive alternative to LEED. Green Globes' focus on modeling energy-efficiency after occupancy is also favored over a sometimes burdensome commissioning system imposed by LEED requirements.

U.S. ENVIRONMENTAL PROTECTION AGENCY ENERGY STAR PROGRAM

The Energy Star program developed by the U.S. Environmental Protection Agency (EPA) is a national energy performance rating system that benchmarks the energy performance of a wide range of commercial facilities relative to the performance of similar facilities across the United States. To be eligible to receive a rating from the Energy Star program, at least 50 percent of a building's floor area must be defined by one of the eligible space types, which assigns the building to a peer group against which the facility will be compared. Based on their space type, geographical location, and level of business activity, the program assigns each facility a national energy performance rating on a scale of 1 to 100. Facilities that meet certain criteria and achieve a rating of 75 or better are eligible to apply for an Energy Star designation.

Studies by the EPA show that the more than 3,200 buildings nationwide that have earned the Energy Star rating use about 35 percent less energy than comparable buildings.

NATIONAL INSTITUTE OF BUILDING SCIENCES WHOLE BUILDING DESIGN GUIDE

The Whole Building Design Guide (WBDG) is a diverse collection of sustainable building resources openly available on the Internet at www.wbdg.org. The WBDG provides links to numerous other Internet resources dealing with practically every aspect of building construction and operation.

Payback and Return on Investment

Green savings are commonly expressed as either a payback period or return on investment (ROI). Either form is acceptable as long as the underlying assumptions are correct. ROI is the more commonly accepted method of evaluating business investments, and perhaps the more useful indicator since it translates directly onto the balance sheet. Payback period (represented in months or years until the cost of an energy-efficiency investment is recovered) is generally unimpressive in business terms, both conceptually and numerically. A two-year payback period certainly sounds less desirable than a 50 percent annual ROI. Moreover, it is not really beneficial in assessing the worthiness of an investment. The ROI may be expressed in one of the following four forms:

Simple Payback Period (PP): This is the most basic means of quickly assessing a proposed improvement's value and is normally used for residential or small-scale improvements. PP is represented as time in years necessary to recover the improvement cost. It is calculated as follows:

$$PP = \text{Cost of Improvement/Annual Savings}$$

Simple ROI: This is a simple form of assessing ROI since it does not consider factors like time value of money, cost of financing, or inflation. This is the method used by many of the online calculators created for various types of projects. ROI is represented as an annual percentage return. It is calculated as follows:

$$ROI = (\text{Gain from Improvement} - \text{Cost of Improvement})/\text{Cost of Improvement}$$

Net Present Value (NPV): NPV is calculated by adding together the monetary benefits of an energy-efficiency measure over a period of its estimated life and subtracting the costs there from. NPV is a better means than the internal rate of return (IRR) for comparing a number of alternative energy-efficiency improvements against one another. If NPV > 0, the project should be considered. If NPV = 0, the project will neither gain nor lose money for the business. If NPV < 0, the project should be rejected, unless there are human resources or other factors that make it beneficial for the business.

Internal Rate of Return (IRR): This value is useful as a comparison against the rates of return or interest rates of alternative monetary investments available to the business, including stocks and bonds. Energy costs are normally tax deductible for businesses, so the IRR should be compared against the after-tax rate of return for taxable investments. Benefits and costs must be discounted in an IRR calculation to account for the time value of money (or the ability to earn interest or appreciation on an investment).

Calculating NPV or IRR for energy-efficiency improvements requires the following information and assumptions:

- Calculate the initial cost of the improvement ($).
- Assumed life of the improvement (years).
- Estimate of first-year energy savings ($).
- Estimate of the annual increase in energy cost (% per year).
- Additional or reduced operating cost resulting from improvement ($ per year).
- Assumed operating cost inflation factor (% per year).
- Cost of financing the improvement (% per year).

Refer to McGraw-Hill's online resources of this book for NPV and IRR calculators. Online NPV and IRR calculators can also be found at http://www.energytools.com/calc/EnerEcon.html.[10]

The *ROI Quik-Calc* boxes included in this book are provided as a quick and rough tool for managers to use in making initial assessments about the benefits of various facility improvements. They should be considered a starting point only, a predecessor to a more detailed and project-specific cost and benefit analysis. See Fig. 1.2 for a summary of the methods of assessing ROI.

There is one significant caveat to using these calculations as a sole means to assess green initiatives. The benefits of some sustainability proposals cannot be easily measured in financial terms. Employee retention and loyalty resulting from improved indoor air quality or a company carpool/ride-share program, for instance, can be of substantial value to a company even though it is difficult to quantify that value. Conversely, HVAC modifications that shorten operating hours or adjust set points may yield significant savings but they may also generate employee dissatisfaction and reduce productivity. Managers must weigh tangible and intangible factors in assessing which sustainability measures to implement in their facilities.

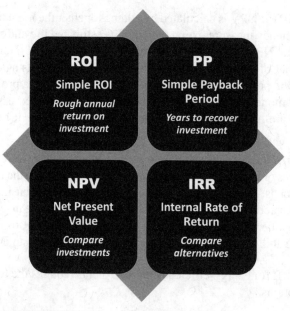

Figure 1.2   Methods of assessing return on investment (ROI).

Energy Calculators and Software

There is no shortage of online calculators, offered by government and industry, to assess the savings associated with energy-efficient improvements. These calculators typically offer basic dollar per year savings for standard improvements, and can be useful to a manager in making a preliminary assessment as to whether a particular type of improvement is worth examining in greater detail. The calculators are most appropriate for office improvements (lighting, plumbing, standard HVAC), and some can help assess construction waste recycling savings or office waste reduction plans. All calculators should be used for initial assessment purposes only, as a way to identify the most promising strategies for a particular site. They should be followed up with detailed estimates based on each facility's specific costs and requirements. See Chap. 8 for a more extensive list of software. See Table 1.6 for a list of major government and organizational sources of sustainability calculators and other tools.

DOE's Office of Energy Efficiency and Renewable Energy   It provides calculators and tools aimed largely at the residential market, but they can be useful for multi-family landlords or small businesses in assessing insulation upgrades. http://www1.eere.energy.gov/consumer/calculators/homes.html

Tip Box

TABLE 1.6   MAJOR GOVERNMENT AND ORGANIZATION SOURCES FOR SUSTAINABILITY TOOLS

U.S. Department of Energy (DOE)
www.energy.gov

U.S. DOE Energy Efficiency and Renewable Energy Department (EERE)
www.eere.energy.gov

U.S. DOE EERE Federal Energy Management Program
www1.eere.energy.gov/femp

U.S. DOE EERE Industrial Technologies Program
www1.eere.energy.gov/industry

U.S. DOE Energy Information Administration
www.eia.doe.gov/pages

U.S. DOE Office of Environmental Management
www.em.doe.gov

U.S. Environmental Protection Agency
www.epa.gov

National Institute of Building Sciences Whole Building Design Guide
www.wbdg.org

Athena Institute
www.athenasmi.org

Envirolink
www.envirolink.org

Healthy Building Network
www.healthybuilding.net

U.S. Green Building Council
www.usgbc.org

WaterSense calculator   The EPA has created a calculator to estimate the savings resulting from installing WaterSense-labeled fixtures and fittings. http://www.epa.gov/watersense/calculator/index.htm

Construction waste estimator   Peaks to Prairies' Pollution Prevention Information Center, funded by the EPA Region 8, provides a tool that can be used to estimate how much waste will be created during a construction project and the disposal costs required to dispose of this waste, enabling a company to evaluate its potential for reduction, reuse, or recycling of construction waste. http://peakstoprairies.org/p2bande/construction/C%26DWaste/calculate.cfm

EPA's WasteWise program tools   WasteWise is an EPA program that helps its partners meet goals to reduce and recycle municipal solid waste and selected industrial wastes. Businesses of all sizes and from all industry sectors can join WasteWise. It provides several tools, described below, to help them in their waste reduction efforts. http://www.epa.gov/epawaste/partnerships/wastewise/index.htm.

Measuring Your Progress   It is a template that can be used by a business to assess its green program achievements. It provides guidelines and resources for locating data sources, calculating waste reduction results, and determining the environmental and economic benefits of a particular program. http://wastewise.tms.icfi.com/measure.htm.

The Waste Reduction and Buy Recycled Tracking Sheet   It allows businesses to track their waste prevention efforts, recycling, and purchase of recycled products for one year. It includes five worksheets: Waste Material, Waste Prevention, Recycling, Buy Recycled, and Summary. Companies evaluate their program by comparing this information to baseline data or results from the previous year. http://wastewise.tms.icfi.com/measure/tracking.htm.

The Measure of Success—Calculating Waste Reduction Booklet   The booklet covers the benefits of measuring waste reduction. It provides step-by-step instructions on how to establish or improve a business waste measurement system and explains a variety of options for different levels of effort and expense. http://epa.gov/epawaste/partnerships/wastewise/pubs/wwupda11.pdf.

U.S. Department of Energy Vehicle Calculator   The U.S. DOE Office of Energy Efficiency and Renewable Energy's vehicle calculator helps calculate the potential savings from converting to a more fuel-efficient vehicle. http://www1.eere.energy.gov/consumer/calculators/vehicles.html.

WAste Reduction Model (WARM)   EPA created the WAste Reduction Model (WARM) to help solid waste planners and organizations track and report reductions in greenhouse gas emissions as a result of different waste management practices. WARM calculates greenhouse gas emissions of baseline and alternative waste management practices, including source reduction, recycling, combustion, composting, and landfilling. WARM is available both as a Web-based calculator: http://epa.gov/climatechange/wycd/waste/calculators/Warm_Form.html and as a Microsoft Excel spreadsheet: http://epa.gov/climatechange/wycd/waste/calculators/downloads/WARM.zip http://epa.gov/climatechange/wycd/waste/calculators/Warm_home.html.

PURCHASING AND PROCUREMENT

U.S. EPA purchasing calculator   The purchasing calculator: http://www.energystar.gov/index.cfm?c=bulk_purchasing.bus_purchasing is a tool designed to assist businesses in making smarter purchasing decisions. Categories include residential, commercial, consumer, and office products.

Xerox office equipment calculator For a Xerox sustainability calculator for printers and office copiers, visit http://www.consulting.xerox.com/flash/thoughtleaders/suscalc/xeroxCalc.html

Endnotes

1. Stika, Nicole. Council of Smaller Enterprises. "Sustainability/Green Plus." http://www.cose.org/Member%20Benefits/Business%20Savings/Energy%20Solutions/Sustainability.aspx. Accessed February 18, 2010.
2. Rheault, Magali. Gallup.com. "In Top Polluting Nations, Efforts to Live 'Green' Vary." http://www.gallup.com/poll/106648/top-polluting-nations-efforts-live-green-vary.aspx. Accessed March 7, 2010.
3. Environmental Leader. "For Hospitals, Some Sustainability Changes Pay Off in 2 Years." http://www.environmentalleader.com/2009/07/28/for-hospitals-some-sustainability-changes-pay-off-in-2-years/. Accessed January 10, 2010.
4. Albinson, Tim. Sourcing Innovation. "A Lesson from the Leaders: Sustainability is the Key to Savings." http://blog.sourcinginnovation.com/2009/11/23/a-lesson-from-the-leaders-sustainability-is-the-key-to-savings.aspx?ref=rss. Accessed February 18, 2010.
5. Timberland.com. "CSR-Environmental Stewardship." http://www.timberland.com/corp/index.jsp?page=csr_climate_impact. Accessed January 10, 2010.
6. Walmart.com. "Fact Sheets." http://walmartstores.com/pressroom/factsheets/. Accessed January 10, 2010.
7. Green, Cindy. The Green Pages. "Small Business Sustainability: It Can Be Done." http://thegreenpages.ca/portal/bc/2008/07/small_business_sustainability.html. Accessed January 11, 2010.
8. Society for Human Resource Management. "Green Company Programs Increase in U.S." http://www.shrm.org/Publications/HRNews/Pages/GreenProgramsIncrease.aspx. Accessed January 12, 2010.
9. Triple Pundit (Environmental News Network). "Is LEED green enough? Conversations from Dwell on Design LA 2008." http://www.enn.com/lifestyle/article/37354. Accessed January 10, 2010.
10. Energy Tools.com. "Economics of Energy Efficiency." http://www.energytools.com/calc/EnerEcon.html. Accessed March 7, 2010.

2

INDOOR HEALTH AND
THERMAL COMFORT

Managing thermal comfort and issues of indoor health are among the most challenging problems faced by business managers. These two areas most affect employee productivity, absenteeism, and satisfaction with their employer. Although studies vary on the degree to which worker comfort affects their productivity, there is little debate that employees who must occupy a space they find thermally uncomfortable are not as productive as they can be. More importantly, employees who believe their workplace is a factor in their health problems are more prone to illness, less productive, and a potential liability threat for an employer.

Balancing economy and effectiveness in managing the indoor environment is a tall order, and one fraught with contradictions for a manager. Most employees, when given the choice, would prefer to have operable windows in their workplace for access to outdoor air when seasonal conditions make it useful for cooling or ventilating a building. Heating, ventilating and air conditioning (HVAC) engineers routinely design building mechanical systems with air intake economizers to take advantage of these opportunities. However, allowing employees to exercise the freedom of simply opening windows can play havoc with a facility's sophisticated air balancing and filtering systems. An employee who relishes connection with the outdoors may sit next to another employee who suffers from asthma complications whenever open windows admit pollen and pollutants.

Thermal comfort issues can be equally bedeviling. Large, central HVAC units offer equipment and installation economies, but deny local occupant control and may preclude shutting down zones of a facility that are not in use. Managers who attempt to realize utility savings by raising or lowering thermostat settings often find some of the savings taken back through employees using under-desk space heaters or other measures to restore comfort.

It would seem, therefore, that the first lesson of sustainability in the workplace is for an employer to know his people well. The cold-natured employees should be kept warm. The asthma sufferers should be provided with an environment that does not contribute to their suffering. After safety and comfort, there will be economies in operation for an employer to take advantage of. But sustainability only works in the workplace when

TABLE 2.1   COMMERCIAL HVAC UNIT EFFICIENCY MEASUREMENT

EQUIPMENT TYPE	EFFICIENCY RATING	EFFICIENCY MEASUREMENT
Gas furnace	Annual Fuel Utilization Efficiency (AFUE)	BTUs of heating output divided by the BTUs of fuel input during a normal heating season.
Gas boiler	Annual Fuel Utilization Efficiency (AFUE)	
Oil furnace	Annual Fuel Utilization Efficiency (AFUE)	
Oil burner	Annual Fuel Utilization Efficiency (AFUE)	
Electric chiller	Integrated Part Load Value (IPLV)	Average kW of input power per ton (12,000 BTU/h) of cooling output.
Heat pump (split)—cooling	Seasonal Energy Efficiency Ratio (SEER)	Total BTUs of cooling delivered divided by total watt-hours of power used during a normal cooling season.
Heat pump (split)—heating	Heating Season Performance Factor (HSPF)	Total BTUs of heating delivered, divided by the total watt-hours of power used during a normal heating season.
Heat pump (single package)—cooling	Seasonal Energy Efficiency Ratio (SEER)	
Heat pump (single package)—heating	Heating Season Performance Factor (HSPF)	
Window air conditioner	Energy Efficiency Ratio (EER)	Cooling output in BTUs per hour for a watt of input power.
Central air conditioner	Seasonal Energy Efficiency Ratio (SEER)	

employees do not equate it with suffering. See Table 2.1 for a listing of means of assessing HVAC equipment efficiencies.

Heating

Space heating, hot water heat, and process heat are the three generic areas of heating required for commercial and industrial facilities. Energy-efficiency strategies for each of these needs varies with the type and size of system. Generically, space heating and hot water strategies rely on increasing efficiency or reducing load. Process systems rely mostly on enhancing efficiency, since the load required for maintaining productivity does not always offer much opportunity for reduction.

SPACE HEATING

Boilers produce hot water or steam for large commercial or industrial facilities. In contrast to the old image of a huge single boiler rattling away in the boiler room, today's boilers are more compact, less costly, and usually installed as smaller units working in tandem to meet a facility's heating needs. Modern boilers feature improved fuel burning and combustion control equipment, and programmable microprocessor-based flame control systems.

Warm air furnaces fire natural gas through curved metal tubes called heat exchangers. Air passing over the heat exchanger absorbs heat from the hot surface of the tube, from where it then travels through the building's ductwork to heat interior spaces. Exhaust gases from the heat exchanger are released outside through vent pipes to the exterior, though some systems recapture part of the exhaust heat to improve the system's efficiency.

The most common type of warm air furnace found in commercial facilities is the rooftop system. These systems are typically horizontal flow units that are built as insulated, weather-tight units for outdoor use. Rooftop units can be designed to provide heating only, or can be fitted to provide all the heating and cooling needs of a facility.

PTACs (Packaged Terminal Air Conditioners) are individual, locally controlled heating and cooling systems. PTACs offer low initial cost, simplified installation, and work well for small building areas such as office condominiums, apartments, or hotel/motel units. Each unit contains all the heating and cooling components in one system, and is designed for thru-wall installation with a remote condensing unit and exterior air intake/exhaust louver. With a PTAC, each space is an individual occupant-controlled zone, simplifying installation and providing the benefit of separately controlled heating and cooling of the space by its occupants. PTAC installations eliminate expensive ductwork, pumps, and central boilers. For large facilities, however, PTAC systems can become expensive compared to central plant systems and do not offer managers the ability to monitor and control the entire facility's HVAC systems efficiently. See Table 2.2 to estimate the return on investment from replacing an existing central HVAC unit with PTAC units.

Infrared heating systems use radiant energy to heat objects instead of air. The system's gas burner produces hot radiative energy from emitter tubes or ceramic surfaces. Optional reflectors can be used to direct radiant heat to select areas. Heaters can be pole-, wall-, or ceiling-mounted, and used indoors and outdoors. Integrated temperature controls for each unit regulate the heaters in a designed system. Low-intensity infrared heaters are ideal for interior or exterior heating applications requiring draft-free directional heat. See Fig. 2.1 for a breakdown of the types of space-heating systems used in commercial facilities.

High-intensity infrared heaters are best for spot heating applications. They can also be used in series for a complete radiant heating system.

ROI Quik-Calc

TABLE 2.2   PACKAGED TERMINAL AIR CONDITIONER (PTAC) REPLACEMENT

The equipment and installation costs for PTAC systems are more than for central air conditioning systems. Use this rough rule of thumb to estimate the savings from replacing an existing central HVAC unit with individual PTAC units.

Multiply number of PTACs to be installed × $4000 installation cost = PTCost (adjust for local and project-specific costs)

For motel/hotel or office occupancies, multiply number of PTAC units × $60 per month to estimate annual savings (AS).

For apartment and condominium occupancies, multiply number of PTAC units × $40 per month to calculate AS.

Divide PTCost (total installation cost) by AS to estimate rough ROI (return on investment).

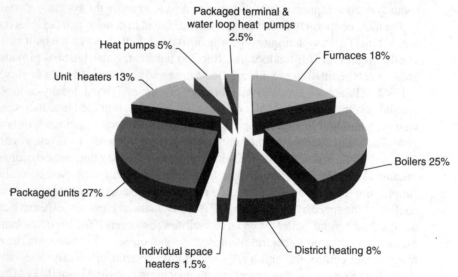

Figure 2.1   Commercial heating unit types.

Because of their unique heating characteristics, infrared heaters are ideal for hard-to-heat areas including:

- Entry vestibules or entryways
- Outdoor areas
- Loading docks
- High-ceiling buildings
- Facilities with rapid temperature changes

Tip Box

TABLE 2.3 HVAC AIR-COOLED EFFICIENCY RECOMMENDATIONS

PRODUCT TYPE AND SIZE[a]	RECOMMENDED LEVEL	BEST AVAILABLE
<65 MBtu/h (3 phase)	12.0 SEER or more[b]	14.5 SEER
65–135 MBtu/h	11.0 EER or more	11.8 EER
	11.4 IPLV or more	13.0 IPLV
>135–240 MBtu/h	10.8 EER or more	11.5 EER
	11.2 IPLV or more	13.3 IPLV

[a]Only air-cooled single-packaged and split system units used in commercial buildings are covered. Water source units are not covered by ENERGY STAR® but look for efficiency ratings that meet or exceed these levels for air source units.

[b]When operating conditions are often close to rated conditions or in regions where there are high demand costs, look for units with the highest EER ratings that also meet or exceed this SEER.

EER (energy-efficiency ratio) is the cooling capacity (in Btu/h) of the unit divided by its electrical input (in watts) at the Air Conditioning and Refrigeration Institute's (ARI) standard peak rating condition of 95°F.

SEER (seasonal energy-efficiency ratio) and IPLV (integrated part-load value) are similar to EER but weigh performance at different (peak and off-peak) conditions during the cooling season.

Source: US Department of Energy, Energy Efficiency and Renewable Energy Program

http://www1.eere.energy.gov/femp/procurement/eep_unitary_ac.html

Thru-wall combination heating/cooling systems work well in limited spaces. The units are self-contained, prewired, precharged systems, and do not need an external condensing unit. Thru-wall systems are low-maintenance and easy to service. They can fit in a small closet, and provide occupants with complete control for the limited space they serve. Depending on the quality and capability of the unit, fan settings may be limited or not. The units provide generally even heating and cooling within a small space, though the ability to heat the far reaches of a room is limited in spaces that tax their maximum capabilities. See Table 2.3 for recommended HVAC system efficiencies.

HOT WATER

Though hot water usage for nonprocess needs may be limited in many commercial buildings (toilets and cafeterias), water heating still accounts for a significant portion of energy used at many facilities. Nationwide, the Whole Building Design Guide estimates that approximately 4 percent of energy used in commercial buildings is for water heating.[1] Solar-powered water heating systems, which use solar energy rather than electricity or gas to heat water, can provide up to 80 percent of the hot water needs of commercial facilities, largely eliminating pollution and fuel costs associated with this need, as well as saving the operation and maintenance expenses of hot water heaters. The savings opportunities for installing solar hot water heaters vary with the heating load of a particular facility, and the contractors involved in this sector are dominated by those with mostly residential experience. Including residential applications, in 2003, solar hot water heating systems provided only 1 percent of the total water heating market (and much of this was for residential swimming pool heaters).

Tankless hot water heaters are a good solution for office or plant toilets, where hot water usage may be heavy at particular times of the day (shift changes and lunch hours, for example), but very light for other extended periods. Rather than store hot water at a constant temperature at all times, tankless units produce hot water on demand via high-efficiency heaters. Although more expensive to install than tank-type hot water heaters, tankless heaters may be a better value in a commercial environment due to lower operation and maintenance costs (depending on use). See Table 2.4 to estimate the return on investment for a tankless versus a tank-type hot water heater.

Solar water-heating systems are most effective at facilities that have a mostly south-facing roof, or unshaded ground area that permits an unobstructed view of the southern sky. The Whole Building Design Guide rates solar-powered hot water systems as most effective when the following characteristics are present[2]:

- The water heating load is largely constant throughout the year (not vacant in summer).
- The water heating load is constant throughout the week (routine shift work).
- The cost of fuel used to heat water is high in the facility's area (more than $10/MBtu), or is limited to electricity or propane.
- A sunny climate is desirable but not essential. In 2003, the three largest markets in the United States for solar water heating were Florida, California, and New Jersey.

ROI Quik-Calc

TABLE 2.4   TANKLESS VS. TANK-TYPE HOT WATER HEATER ROI

TANK-TYPE HOT WATER HEATERS

- Installation: Multiply the number of hot water heater locations × $500 average installation cost.

- Operation: Multiply the number of hot water heater locations × $750 annual operating cost. Multiply this total × 10 years.

Add the installation and 10-year operating cost.

TANKLESS HOT WATER HEATERS

- Installation: Multiply the number of hot water heater locations × $800 installation cost.

- Operation: Multiply the number of hot water heater locations × $650 annual operating cost. Multiply this total × 10 years.

Add the installation and 10-year operating cost.

Compare total 10-year costs. For single heater installation, using the above estimates would result in an ROI of four years and a savings of $3700 over a 10-year period.

Note: Tankless water heaters are estimated to have at least 50% longer life span than tank-type heaters.

Cooling

UNITARY OR PACKAGED AIR CONDITIONING SYSTEMS

Unitary (also referred to as packaged or DX) systems are the most widely used type of air conditioning equipment in the United States. Unitary systems are often roof-mounted, and provide cooling through the refrigerant vapor-compression cycle. The vapor-compression cycle converts a liquid refrigerant to a gas, and then converts it back again to a liquid, and in the process creates both cooling and waste heat.

Smaller unitary systems utilize a single compressor, and resemble residential units. Multiple compressors are used in systems over 10 tons (120,000 Btu/h) and larger. Multiple compressors have the ability to provide output at various levels, allowing them to cool a space more efficiently when the cooling load is reduced below the maximum capacity. Larger roof-top units may be designed for multizone variable air volume applications and sophisticated control systems, offering advantages typical of chiller-based systems. See Fig. 2.2 for the percentages of cooling system types used in commercial buildings.

SPLIT SYSTEMS

Split systems are unitary systems designed as dual factory-made assemblies. A typical split system may consist of a fan, an evaporative coil, filters, and dampers in an indoor unit, while the compressor and condenser coil are housed in an outdoor cabinet. For split systems, heating is provided by a furnace section or heat pump. A single split system is used for each temperature zone of the building.

Almost all commercial unitary equipment is air cooled. Water-cooled heat pump systems exist, however, and have the capability to handle the cooling loads for large multizone buildings. Packaged water-cooled air conditioner/heat pump combinations provide

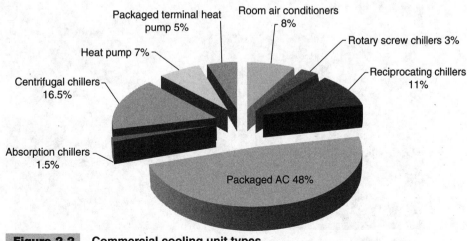

Figure 2.2   Commercial cooling unit types.

an easy way to get multiple uses of a single water loop connecting these units, allowing them to serve multiple zones with different needs, including where interior zones call for cooling while exterior zones require heating. One aspect of large systems of this type is the need to provide a central boiler and cooling tower to supplement the capacity when the demands of the system exceed the loop capacity. Ground water or a ground source heat exchanger can also be used to efficiently provide a lower loop temperature.

One large disadvantage of using a water-source system is that a separate ventilation system must be installed to meet ventilation requirements. These systems also have virtually no ability to work with economizers to take advantage of favorable outside conditions.

Ventilation and Air Movement

Ventilation needs can vary dramatically in a commercial facility, depending on the type of occupancy and the nature of the facility. Industrial processes may require substantial air exhaust to remove contaminants or irritants generated by manufacturing processes. Even more mundane commercial occupancies, such as day care or retail uses, however, may require significant exhaust capabilities as well, particularly where paint, food preparation, or other activities occur. See Fig. 2.3 for a photograph of a typical production area ventilation system.

Historically, HVAC systems have provided ventilation based on assumed or design occupancy, rather than based on how many people are actually in the room at a given

Figure 2.3   Indoor ventilation system.

time. This is a hugely inefficient method, since the system is running as if the facility is fully occupied or otherwise conforming to the design assumptions. Commercial facilities are dynamic occupancies, with constantly shifting human and equipment loads.

A Demand Control Ventilation (DCV) system with a carbon dioxide (CO_2) sensor can provide an accurate reading of a room's actual occupancy level, adjusting the amount of fresh air introduced into the space based on actual need as determined by output from the CO_2 sensor. This type of system offers enhanced indoor air quality and energy efficiency.

NATURAL VENTILATION

In some parts of the country, where temperature and humidity levels permit, natural ventilation through operable windows can be used as an energy-efficient means of supplementing the HVAC systems to provide outside air ventilation and cooling (see the Air Intake Economizer section). Use of ambient air directly, outside of the HVAC system, is also possible, though with some caveats for building operators. Concerns with using outside ambient air include temperature fluctuations, humidity, outdoor air pollution levels, and precipitation. Operable windows can enhance occupants' sense of well-being, connectedness to the outdoors, and feeling of control over their environment. Operable windows can also allow for supplemental exhaust ventilation during construction, maintenance, or activities that may introduce pollutants into the building's indoor environment.

However, sealed buildings with properly designed and maintained HVAC systems may provide better indoor air quality than a building with operable windows controlled randomly by its users. Uncontrolled ventilation with outdoor air can allow outdoor air contaminants to bypass filters, potentially disrupt the balance of the mechanical ventilation equipment, and permit the introduction of excess moisture if access is not controlled. The ability to use outdoor air for ventilation and cooling is also heavily dependent on local environmental factors, such as temperature, humidity, precipitation, and pollution levels. Managers must also consider the effect on their employees (and resultant liability) resulting from pollen, smoke, or other outdoor pollutants that may enter the building through open vents.

Strategies using natural ventilation include wind-driven cross ventilation (horizontal) and stack ventilation (vertical) that use the difference in air densities to provide air movement across a space. Both types of natural ventilation require careful engineering to ensure effective convective flows. The proper sizing and placement of openings is critical, and the flow of air from entry to exit must be planned to avoid obstructions that would render the system useless for large area ventilation.

If operable windows will be used to supplement the HVAC system, ensure that:

- Openings for outdoor air are located between 3 and 6 feet (91.4 and 182.9 centimeters) from the floor (head height).
- The windows are adjustable and can be closed tightly to minimize infiltration or leakage.
- The windows are placed to take maximum advantage of wind direction, with openings on opposite sides of the building to maximize cross ventilation.

Figure 2.4   HVLS fan.

INDUSTRIAL CEILING FANS

Large-diameter industrial ceiling fans are called HVLS (high volume, low speed) fans. Fans such as those manufactured by Big Ass Fans, Inc. (Lexington, KY) can be ordered in diameters as large as 24 feet (7.32 meters). The company's Powerfoil X Plus model can displace over 100,000 cfm of air at 84 rpm, providing an energy-efficient means of supplying general ventilation in industrial spaces. Their largest fan can supply ventilation for a 30,000-square-foot (2787-square-meter) facility.[3]

See Fig. 2.4 for a photograph of an HVLS fan.

MacroAir Technologies produces an HVLS powered entirely by photovoltaic solar energy. The company does not currently offer a battery backup, so the fan's operation is limited to sunlit periods during the day. They do plan to offer a battery backup system in the near future.

Indoor Air Quality

Indoor air quality should be a serious concern for any building or business manager. Increased liability resulting from sick building syndrome, employee sensitivity to volatile organic compounds in building products and finishes, and combustion or other gases resulting from manufacturing processes have increased dramatically in recent

Tip Box

TABLE 2.5   INDOOR AIR QUALITY ISSUES

- Occupant Issues
  - Asthma and allergies
  - Environmental tobacco smoke
- Radon
- Cleaning and Maintenance Products
  - Solvents
  - Paints and coatings
  - Sprays
  - Pesticides
  - Air fresheners
  - Cleaners and disinfectants
- Building Construction and Renovation Activities
  - Formaldehyde
  - Lead-based paint
  - Asbestos
  - Volatile organic compounds
  - Fiber and dust generation
  - Polychlorinated biphenyl (PCB) contamination
- Biological
  - Mold
  - Dust mites
  - Pests (cockroaches and rodents)
  - Pets
- Combustion Pollutants
  - Gases
  - Combustion particulates
  - Carbon monoxide
  - Nitrogen dioxide
  - Sulfur dioxide
  - Excess water vapor

years. Employees spend 9–10 hours per day in their workspace and are demanding, rightfully so, that employers take responsibility for ensuring that the indoor air quality of their environment is safe. See Table 2.5 for a list of common indoor air quality issues.

ASTHMA

Asthma is a life-threatening respiratory illness that affects over 20 million Americans. It is a more serious workplace liability and productivity problem than is widely recognized by employers. According to the U.S. Environmental Protection Agency (EPA), asthma affliction rates have risen dramatically over the past thirty years, particularly among children aged 5 to 14.[4]

Principal workplace triggers of asthma are as follows:

- Secondhand smoke results from the smoke from the burning end of a cigarette, pipe, or cigar, or from the smoke exhaled by a smoker in an indoor or outdoor setting.
- Dust mites are too small to be seen, and are most often found in home mattresses. In commercial settings, however, they may also be present in carpets, upholstered furniture, and curtains.
- Mold can grow indoors when mold spores land on wet or damp surfaces, particularly where they have a ready source of paper or other material to consume. In commercial facilities, mold is most often found in service areas, kitchens, toilets, water/electric meter/fire pump rooms, and other areas of dampness or exterior wall penetrations.
- Cockroaches and other pest body parts, secretions and droppings, as well as the urine, and saliva of pests, such as rodents, are often found in areas of commercial facilities where food and water are present.
- Hair and fur from warm-blooded pets, such as dogs and cats, either kept as pets or mascots at a facility, or those who infiltrate the building surreptitiously.
- Nitrogen dioxide is a reddish-brown, irritating-odor gas that results from the by-product of indoor fuel-burning appliances, such as gas stoves, gas or oil furnaces, or unit heaters, fireplaces, or wood stoves.

Effectively managing a workplace environment for asthma sufferers can best be accomplished by working with the employee (in consultation with his or her physician) to develop a plan that includes the use of medications in combination with the avoidance of environmental triggers. This is particularly important in school, day care, and institutional settings (such as assisted living or healthcare facilities). Since children may spend much of their time in schools and day care facilities, it is important to reduce their exposure to environmental asthma triggers as much as possible in each of these environments. The U.S. Green Building Council (USGBC) addresses indoor environmental quality in their LEED Existing Buildings certification under the Environmental Quality (EQ) category. See Table 2.6 for a summary of these requirements.

TABLE 2.6   LEED EXISTING BUILDING ENVIRONMENTAL QUALITY (EQ) CATEGORIES

- EQ Credit 1.2: Outdoor air delivery monitoring
- EQ Credit 3.2: Green cleaning—Custodial effectiveness assessment < score of 3
- EQ Credit 3.3: Green cleaning—Custodial effectiveness assessment < score of 2
- EQ Credit 3.4: Use of 30% sustainable cleaning products and materials
- EQ Credit 3.5: Use of 60% sustainable cleaning products and materials
- EQ Credit 3.6: Use of 90% sustainable cleaning products and materials
- EQ Credit 3.7: Use of sustainable cleaning equipment

Common strategies for controlling asthma in the work environment include:

Animal dander Ban domestic pets from the school or workplace. If present, keep them out of areas where children or those with asthma issues spend most of their time. Restrict pets from carpeted areas.

Cockroaches To eliminate the potential for cockroach problems, it is essential that water and food sources be controlled. Restrict employee food consumption to a cafeteria or other areas (no desktop consumption), insist that food be stored in tightly sealed containers, and make employees responsible for cleaning the kitchen/eating area themselves (not relying on cleaning service). Create a culture of never leaving food, water, or garbage exposed.

Dust mites Dust mites live in carpets, pillows, drapes, and fabric-covered furniture. Put pillows and fabric-covered furniture in dust-proof (allergen-impermeable) zippered covers that can be removed and cleaned. Clean drapes periodically (once every six months at least). Ban stuffed toys, novelties, or other items from the workplace, or at least limit the amount of time they can be kept.

Mold Repair workplace water problems and leaks promptly (within 24 to 48 hours). In the event of flooding or water infiltration, thoroughly dry wet areas to prevent mold growth. Discard water-damaged paper or porous items. Allow affected employees to work from home until water-damaged area has been thoroughly dried and tested for mold spores.

Secondhand smoke Do not allow smoking within the building. Insist that smoker employees use a designated outside area, remote from operable windows or air intakes (see Environmental Tobacco Smoke section for additional details on controlling secondhand smoke).

MOLD

Mold exposure represents a significant indoor environmental hazard faced by commercial building managers. When moisture problems occur and mold growth results, building occupants may begin to report odors and a variety of health problems, including: allergic reactions, headaches, skin irritation, breathing difficulties, and aggravation of asthma symptoms. All of these symptoms could potentially be associated with mold exposure. All molds have the potential to cause health effects in certain individuals. Molds produce allergens, irritants, and even toxins that may cause reactions in workers, ranging from mild to severe and even life-threatening. The types and severity of symptoms depend, to various extents, on the types of mold present, the ages of the individuals, the extent of a person's exposure, their underlying health factors, and their existing sensitivities or allergies.

According to the EPA, specific reactions to mold growth can include the following[5]:

Asthma   Molds can trigger asthma attacks in persons who are already sensitive to the particular type of mold. The irritants produced by mold spores may also worsen the effects of asthma even in those individuals who are not allergic to molds.

Allergic reaction   Inhaling or touching mold spores and molds may cause immediate or delayed allergic reactions in individuals who are susceptible. Allergic responses include hay fever–type symptoms, including skin rashes (dermatitis), runny nose, red eyes, and sneezing. Additionally, repeated exposure to mold spores may cause previously nonsensitive individuals to become sensitive, or increase the sensitivity of those with milder initial reactions. Mold spores and fragments can also cause allergic reactions in sensitive individuals regardless of whether the mold is dead or alive.

Irritations   Exposure to mold and mold spores can cause irritation in the eyes, nose, skin, throat, and lungs, often in the form of a burning sensation.

Hypersensitivity pneumonitis   Hypersensitivity pneumonitis is a disease that mimics bacterial pneumonia. Though uncommon, it may develop following either short-term or long-term exposure to molds.

Opportunistic infections   People with compromised immune systems may be more vulnerable to infections by molds. A form of mold known as *Aspergillus fumigatus*, in particular, has been known to infect the lungs of immune-weakened individuals. When susceptible individuals inhale the mold spores, they begin growing in their lungs and overtax their body's diminished immune system. Healthy individuals are not usually vulnerable to opportunistic infections resulting from exposure to airborne mold in the environment. Exceptions always exist, however, so the most prudent course for building managers is to eliminate all known sources of mold in their facilities.

Skin diseases   Some molds can trigger skin diseases, including athlete's foot and yeast infections.

RADON

Radon is an odorless, tasteless, and invisible gas produced by the decay of naturally occurring uranium in soil and water. Radon is a form of ionizing radiation and a proven carcinogen. Lung cancer is the only known effect on human health from exposure to radon in air. Radon is the leading cause of lung cancer among nonsmokers, according to EPA estimates.[6] Overall, radon is the second leading cause of lung cancer in the United States, responsible for approximately 21,000 lung cancer deaths every year. This total includes 2,900 deaths among people who have never smoked. Radon is primarily a problem in the home environment, where fewer air changes and longer periods of exposure increase the risk. Commercial facility managers must not ignore this hazard, however, since it can have long-term and severe consequences for individuals who are exposed over a period of years to radon gases.

Radon in outdoor air is common to a limited extent, and is present in building indoor air in greater concentrations depending on the area where the facility is located, whether above or below grade space, that type of foundation, and the relative tightness of the construction. EPA recommends that residences be remediated if the radon level is at or above 4 pCi/L (picocuries per liter). But because there is no known safe level of exposure to radon, EPA also recommends that Americans consider repairing their home for radon levels above 2 pCi/L. Building managers should test their facilities for radon exposure once every five years and follow these same guidelines.

CLEANING AND MAINTENANCE PRODUCTS

Building cleaning products can produce harmful effects to occupants who are sensitive to chemicals used in the products, when cleaning services or employees use excessive amounts of products, or when the ventilation in a particular area is inadequate to flush the air of contaminants. LEED Existing Building certification requires the use of green cleaning products and equipment under the EQ category.

Building managers must pay particular attention to those products containing volatile organic compounds (VOCs), which are organic solvents that easily evaporate into the air. Although some VOCs may be flammable, the larger concern is employee susceptibility to a product's components. Potentially hazardous compounds (required to be listed on product labels) include: petroleum distillates, mineral spirits, chlorinated solvents, carbon tetrachloride, methylene chloride, trichloroethane, toluene, and formaldehyde. Exposure to some cleaning products may cause eye, nose, and throat irritation; nausea; dizziness; loss of coordination; or headaches. Some products have also been associated with liver, kidney, or central nervous system damage and cancer. Building managers should maintain a file of Hazardous Materials Data Sheets for all cleaning products used in the building, and restrict employees from bringing products from home that are not approved. See Chapter 4 for information on implementing a green cleaning program in the workplace.

BUILDING EXHAUST SYSTEMS

Robust building exhaust systems integrated into a facility's HVAC operation can be a solution to many indoor air-quality issues. These systems reduce energy-efficiency, however, so should be carefully sized and limited in operation to cure the concern with minimal energy consumption. In particular, additional static pressure beyond the minimum necessary to exhaust the system adds operating costs in terms of excess electric and auxiliary heat consumption. See Table 2.7 for exhaust system location tips.

Tip Box

TABLE 2.7   TIPS FOR EFFICIENT EXHAUST SYSTEMS
■ Locate exhaust intakes close to the source.
■ Place supply outlets close to the worker.
■ Use makeup air blowers year round.
■ Control the flow to match the contamination demand.

COMBUSTION POLLUTANTS

Combustion pollutants in facilities include gases, particles, and excess water vapor produced as a by-product of burning fuels such as natural gas, propane, oil, wood, kerosene, and coal. Harmful combustion gases include carbon monoxide, nitrogen dioxide, and sulfur dioxide. Other harmful by-products are particulates and excess water vapor.

The results of combustion pollutants range from adverse health effects to potential fatalities. Carbon monoxide is an odorless gas that can be fatal, nitrogen dioxide may damage respiratory tracts, while sulfur dioxide can irritate eyes, noses, and respiratory tracts. Smoke and particulates have been associated with lung cancer; can irritate the eyes, noses, and throats of workers; and may worsen chronic respiratory illnesses, such as asthma. Too much water vapor can lead to moisture problems and biological growth in poorly ventilated areas, including the growth of mold.

Combustion pollutants may enter the workplace through a variety of sources. Unvented, improperly installed, or poorly maintained heating, drying, baking, or cooking appliances that burn fuels are the usual culprits. These include a wide range of equipment and appliances found in commercial, industrial, institutional, and office occupancies, and include furnaces, boilers, water heaters, fireplaces, stoves, space heaters, dryers, and ovens.

To avoid combustion gas issues in the workplace, managers must follow a regular safety check and preventative maintenance program for any equipment producing combustion gases. The safety review should include periodic air monitoring, since harmful build-ups of these gases can occur when exhaust from combustion equipment is not properly vented or redirected, when the equipment is not in good working order, or when equipment is not regularly inspected for safe operation.

Managers must also be sensitive to the potential for combustion gas problems on a building basis associated with the problem of backdrafting. Backdrafting occurs when the air pressure inside an occupied area is less than the air pressure outside, or in another part of the facility, potentially causing exhaust gases or other pollutants to be pulled into the habitable part of the facility.

Combustion by-products may also enter the occupied space rather than being vented outside. This hazard can occur when windows or skylights are opened adjacent to exhaust vents, or when plant combustion gases are inadvertently pulled into an office or smaller area through pressure changes or the operation of mechanical ventilation in the occupied area. For other than carbon monoxide, employees can often be the most effective monitor of environmental change of combustion gas entering the workplace, though the prudent manager will rely mostly on periodic testing of air quality to ensure a safe environment for his employees.

Air Intake Economizers

An air intake economizer (also called airside economizer in ASHRAE 90.1) is a duct, damper, and control arrangement that allows a building's mechanical system to use outside air to help cool a building when conditions permit. The use of an air intake economizer system is now mandated by the International Code Council's International Energy Conservation Code (IECC) through its referencing to ASHRAE standards.

When a building gains heat from its occupants, lighting, and equipment, the internal heating load continues to heat the building even if the outside ambient temperature is comfortable.

During these periods it is more economical to shut off the compressors and cool the facility using outdoor air. Integral economizers in an HVAC system measure the outdoor air's temperature and humidity, determining if it is cool and dry enough to keep occupants comfortable. When the exterior conditions meet the requirements, the economizer will tell the system to use outside air for cooling rather than operating the compressors.

Economizers are most effective in warm, dry climates (western United States) and less effective in warm, moist climates, such as those found in the southern tier of the United States. Cool and moist climates may provide some opportunities for economizer cooling in facilities with high indoor occupant or equipment loads.

ASHRAE 90.1 defines four standard modes of building cooling[7]:

1 **Heating mode:** Minimum outdoor intake air is mixed with air to heat building.
2 **Modulated economizer mode:** During cool weather (30–55°F, or −1.1–12.8°C), the building design temperature can be maintained through a mix of outside air and indoor return air to match the cooling capacity to the cooling load.
3 **Integrated economizer mode:** During warm weather (55–75°F, or 12.8–23.9°C), outdoor air can supply some, but not all of the cooling load. Outdoor air is mixed with mechanically conditioned air to meet the demand in the most efficient manner.
4 **Mechanical cooling mode:** Economizer is off. The HVAC system cools the building with minimal outdoor ventilation air.

ECONOMIZER CONSIDERATIONS

Even though an economizer may not be required in a particular building climate zone, it may make sense for particular types of buildings. Those with high internal heat loads (lots of people using heat-producing equipment) are prime examples. An engineering analysis for economizer systems should be performed even when they are not required by code to determine if they can save energy in a particular facility.

Judging an economizer system requires analyzing the initial cost, operating cost, maintenance cost, and space relative-humidity performance. Trane Corporation recommends that designers appropriately weigh the trade-offs by using an economic/performance analysis program to compare: fixed dry bulb, fixed enthalpy, and differential enthalpy.[8]

Systems Control and Monitoring

Control systems for HVAC systems are electronic, computer-linked systems that control the indoor climate in buildings. They are often networked to allow remote management from off-site locations. HVAC control systems may also integrate with the building's fire, security, and lighting controls into one system, in which case coordination with the requirements of the International Building Code (IBC) and International Fire Code (IFC) will also be required. Such systems typically use one or more central controllers to monitor remote terminal unit controllers. They are used in large commercial and industrial buildings to allow central control of numerous HVAC units around the building. Control systems are now considered an essential component of a large building's energy-management system. Indeed, without a relatively sophisticated HVAC control system, it is not possible to achieve the large efficiencies integrated HVAC systems are capable of providing.

A single control system interface can be used on many commercial HVAC systems to control a variety of equipment types and functions. These include attributes such as advanced diagnostics, remote and alarm monitoring and resets, and system setups.

Central controllers and newer terminal unit controllers are programmable with what is called a direct digital control (DDC) code. This code allows the unit to be customized to work with the system and coordinated with all the other HVAC equipment. A DDC program can allow control of equipment time schedules, set points, controllers, alarms, logic, timers, and trend logs. Unit controllers typically have analog and digital inputs that allow measurement of temperature, humidity, or pressure. Analog outputs control the transport medium, whether hot or cold water, or steam. Digital inputs come from contacts from a control device, and analog inputs are typically a voltage or current measurement from a variable (temperature, humidity, velocity, or pressure) sensing device. Digital outputs typically are used to start and stop equipment, while analog outputs typically issue voltage or current signals to control the operation of devices, such as valves, dampers, and motors.

The DDC system is the operating system of the HVAC system. It controls the position of every damper, valve, and module in the overall system. It determines when fans, pumps, and chillers run, and at what capacity they operate.

Common protocols for this type of control are BACnet® and LonTalk®. Equipment with these capabilities can also be installed on rooftop units to let them communicate with a building's automation system.

Special Systems and Strategies

STRATEGIES FOR SPECIAL SITUATIONS[9]

Following are some strategies building managers may use to address a variety of special situations and building systems.

- Reduce HVAC system operation when the building or space is unoccupied.
- Turn HVAC systems off earlier in the day and allow residual heat and cooling to carry the building through the final few hours of occupancy.
- Shut down HVAC in production areas that are unoccupied, between shifts, or not operating for any reason. Maintain minimal HVAC for maintenance purposes.
- If the system uses direct outside combustion air, close outdoor air dampers.

- Eliminate HVAC usage in vestibules and other unoccupied spaces.
- Minimize direct cooling of unoccupied areas by turning off fan coil units and unit heaters and by closing the vent or supply air diffuser. Use ventilation only after hours.
- Turn industrial and production area fans off during nonworking hours.
- Install HVAC night and weekend setback controls for offices or nonproduction areas.
- Adjust thermostat settings for changes in seasons. Automatic settings should be higher during the cooling season and lower during the heating season.
- Reduce HVAC operating hours, particularly during the temperate spring and fall seasons, by allowing building users to utilize outside ventilation and through permitting wider temperature fluctuations.
- Modify the housekeeping schedule to minimize use of HVAC during low-occupancy periods.
- Install controls to vary the hot water temperature based on the outside air ambient temperature.
- Schedule off-hour meetings in a location that does not require HVAC in the entire facility.
- Install separate controls for HVAC zones. Link them with an energy-management system.
- Install local heating/cooling equipment to serve seldom-used areas located far from the center of the HVAC system.
- Use variable speed drives and direct digital controls on the water circulation pumps, motors, and controls.
- Selectively adjust areas that are too hot or too cold rather than modify the system settings.
- Adjust air duct supply diffusers and return air registers for more effective distribution.
- Use operable windows or vents for cool air ventilation during periods when the outdoor temperature and humidity allow.
- Use window coverings such as blinds, curtains, or awnings to reduce heat loss in the winter and heat gain in the summer.
- Use light-colored roofing material and exterior wall coverings with high reflectance to reduce heat gain. Use exterior trees or shade screens to reduce the amount of sun striking the building's exterior.
- Install ceiling fans to reduce air stratification and increase effectiveness of heating and cooling systems.
- Create smaller HVAC zones with separate controls and appropriately sized equipment.
- Reduce unnecessary heating or cooling system operation during unoccupied hours, or for plant shutdowns.
- Allow a fluctuation in temperature, usually in the range of 68 to 70°F (20 to 21.1°C) for heating and 78 to 80°F (25.6 to 26.7°C) for cooling.
- Adjust heating and cooling controls when weather conditions permit or when facilities are unoccupied.
- Adjust air supply from the air-handling unit to match the required space conditioning.
- Eliminate reheating air to control humidity (often air is cooled to the dew point to remove moisture, then is reheated to desired temperature and humidity).

- Install an economizer cycle instead of operating on a fixed minimum airflow supply. An economizer allows the HVAC system to utilize outdoor air by varying the supply airflow according to outdoor air conditions, usually using an outdoor dry bulb temperature sensor or return air enthalpy (enthalpy switchover). Enthalpy switchover is more efficient because it is based on the true heat content of the air.
- Employ methods of recovering equipment and process heat. A heat exchanger transfers heat from one medium to another. Common types of heat exchangers are: rotary, sealed, plate, coil run-around system, and hot oil recovery system.
- Install heat recovery ventilators that exchange between 50 and 70 percent of the energy between the incoming fresh air and the outgoing return (conditioned) air.
- Minimize the amount of air delivered to conditioned spaces. The amount of air delivered to a space is dependent upon its heating and cooling load, delivery temperature, ventilation requirements, and the number of required air changes. On average, the air should change completely in each space every 5 to 10 minutes.
- Reducing airflow will reduce fan horsepower required for the building, though the United States Occupational Safety and Health Administration (OSHA) and local building code requirements for air exchanges must be followed.
- Extend the amount of time for circulation of air by using a fan discharge damper, fan vortex damper (fan inlet), or fan speed changer.

EXHAUST AND MAKE-UP AIR STRATEGIES

- Make-up air depends on the needs of ventilation for personnel; exhaust air from workspaces overcoming infiltration; machine air needs; and federal, state, and local requirements.
- Properly insulate all walls and ceilings.
- Review process temperatures at regular intervals to minimize waste.
- Reduce air volume lost to the exterior by reducing exhaust rates to the minimum.
- Seal ducts that run through unconditioned space (up to 20 percent of conditioned air can be lost in supply duct run).
- Keep doors closed when the air conditioning is running.
- Install thermally efficient windows to minimize cooling and heating loss.
- Insulate forced-air ducts, as well as chilled water, hot water, and steam pipes.
- Rewire fans to operate only when lights are switched on (if the local electrical code allows).
- Check for duct damper leaks; ensure dampers seal tightly.
- Shut off unnecessary exhaust fans and reduce their use whenever possible (shut down during nonworking hours).

ROUTINE MAINTENANCE STRATEGIES

- Inspect to ensure dampers are sealed tightly.
- Clean coil surfaces annually.
- Replace air filters at regular intervals (typically seasonally).
- Ensure doors and windows seal tightly to reduce air infiltration.

- Check fans for any cause that may reduce flow, including lint and dirt.
- Regularly schedule HVAC system inspections (the typical energy savings generated by a tune-up is 10 percent).
- Reduce fan speeds where possible, and adjust belt drives for proper tightness.
- Regularly check the condition of building thermostats, and calibrate every six months.
- Maintain system controls (repair and replace promptly; calibrate frequently).
- Inspect ductwork for leaks and dust/dirt accumulation.
- Repair leaks in any system component promptly.
- Reduce use of hot water pumps in periods of mild weather.
- Check condition of motor, valves, dampers, and linkages.
- Check/maintain steam traps, vacuum systems, and vents in one-pipe steam systems.

COOLING SYSTEM MAINTENANCE STRATEGIES

- Clean the surfaces of the coiling coils, heat exchangers, evaporators, and condensing units regularly so that they are clear of obstructions.
- Adjust the temperature of the cold air supply from air conditioner or heat pump or the cold water supplied by the chiller. A 2 to 3°F (1.1 to 1.7°C) adjustment can bring a 3 to 5 percent energy savings. Test and repair leaks in equipment and refrigerant lines.
- Repair or upgrade inefficient chillers.

GENERAL MAINTENANCE STRATEGIES

- Clean and adjust the boiler or furnace.
- Check the equipment combustion efficiency by measuring the carbon dioxide/oxygen concentrations, as well as the temperature of stack gases. Adjust for optimum efficiency.
- Remove soot from boiler tubes and heat transfer surfaces on a regular basis.
- Install fuel-efficient burners when replacing or upgrading processes.

OPERATIONAL STRATEGIES

- Determine if the hot air or hot water supply can be reduced.
- Check to see if the forced air fan or water circulation pumps remain on for a suitable time period after the heating unit, air conditioner, or chiller is turned off to distribute air remaining in the distribution ducts.
- Install an energy-management system (EMS) for all or major portions of the facility. An EMS is a system designed to optimize and adjust HVAC operations based on environmental conditions, changing uses, and schedules.

FUEL-BURNING EQUIPMENT STRATEGIES

- Install a more efficient burner.
- Install an automatic flue damper to close the flue when the equipment is not in operation.
- In older fire tube boiler, install turbulators to improve heat transfer efficiency.

- Install an automatic combustion control system to monitor the combustion of exit gases and adjust the intake air for large boilers.
- Insulate all hot boiler surfaces to reduce energy losses.
- Install electric ignitions in lieu of standing pilot lights.

THERMOSTAT STRATEGIES

- Install programmable thermostats.
- Lock thermostat to prevent tampering or constant readjustments by employees.
- Check proper location of thermostats to provide balanced space conditioning.
- Locate thermostats distant from equipment that produces heated or cooled air.

BOILER EFFICIENCY STRATEGIES

- Investigate preheating boiler feed water.
- Adjust boiler and air conditioner controls so that boilers do not fire and compressors do not start at the same time but satisfy demand.
- Use boiler condensate hot water to preheat combustion air.

COOLING TOWER STRATEGIES

- Use existing cooling towers to provide chilled water instead of using mechanical refrigeration for part of the year.
- Install cooling tower water meters to measure make-up water usage.

See Table 2.8 for LEED Existing Building requirements relating to cooling towers.

HVAC STRATEGIES

- Install controls on heat pumps (if this equipment has electric resistance heating elements) to minimize electricity use.

TABLE 2.8   LEED EXISTING BUILDING WATER TOWER REQUIREMENT SUMMARY

This is a summary. Refer to www.usgbc.org for full guidelines.

WE CREDIT 4.1: CHEMICAL MANAGEMENT

Develop and implement a water tower management plan that addresses maintenance issues such as staff training, chemical use, biological control, and bleed-off.

Install and maintain upgraded controls to improve water efficiency.

WE CREDIT 4.2: NONPOTABLE WATER SOURCE USE

Use 50% nonpotable makeup water.

Institute a metered measurement program that verifies the nonpotable water use.

■ Employ cool storage strategies to save on electric bills. The concept behind cool storage systems is to operate the system during off-peak electricity hours, using the lower temperatures stored in the building mass to satisfy the air conditioning demand load during peak hours, thereby reducing electric bills.

■ Install a variable air volume (VAV) system with variable speed drives on fan motors. A VAV system is designed to deliver only the volume of air needed for conditioning the actual load.

■ Upgrade equipment to high-efficiency models where possible. Federal law and some state laws require minimum efficiency levels for energy-intensive equipment. Consider purchasing equipment that exceeds the standards, when upgrading or replacing HVAC or plant components. See Table 2.9 for the sources of federal energy-efficiency requirements for various types of HVAC and plant equipment. Inventory potential system components that may require replacement and keep extra stock on hand for immediate use.

■ Increase the duct size to reduce the duct pressure drop fan speed (delivering more air volume at a lower velocity is more efficient than delivering air at a faster rate through smaller ducts).

■ Assess fan load versus installed units. Replace oversized fans with appropriately sized models.

TABLE 2.9   COMMERCIAL EQUIPMENT EFFICIENCY STANDARDS

■ ASHRAE Products (ASHRAE Standard 90.1)

■ Clothes Washers (*Federal Register*, 75 FR 1122, January 8, 2010)

■ Distribution Transformers (*Federal Register*, 72 FR 58190, October 12, 2007)

■ Electric Motors (*Federal Register*, 75 FR 59657, September 28, 2010)

■ Furnaces and Boilers (*Federal Register*, 10 CFR Part 431, October 21, 2004)

■ High-Intensity Discharge Lamps (*Federal Register*, 75 FR 37975, July 1, 2010)

■ Metal Halide Lamp Fixtures (*Energy Independence and Security Act of 2007*)

■ Packaged Terminal Air Conditioners and Heat Pumps (*Federal Register*, 75 FR 4474, January 28, 2010)

■ Refrigerated Beverage Vending Machines (*Federal Register*, 74 FR 44914, August 31, 2009)

■ Refrigeration Equipment (*Federal Register*, 10 CFR Part 431, January 9, 2009)

■ Small Electric Motors (*Federal Register*, 75 FR 17036, April 5, 2010)

■ Unitary Air Conditioners and Heat Pumps (*Federal Register*, 74 FR 36312, July 22, 2009)

■ Walk-In Coolers and Walk-In Freezers (*Energy Policy Conservation Act of 1975*, Pub. L. 94-163)

■ Water Heaters (*Federal Register*, 10 CFR Part 431, October 21, 2004)

■ Install a variable speed, electronically controlled fan motor that can compensate for varying building load conditions.

Waste Heat Recovery Systems

A waste heat recovery system is a heat exchanger that reuses waste heat generated by a manufacturing or industrial process. Waste heat most often occurs in the form of hot flue gases, but may also occur in the form of steam or hot liquid waste water. In such a system, a heat exchanger is the device used to transfer heat from the hot medium. It is commonly used in industrial processes to improve the efficiency and economy of manufacturing processes of all types. It is also widely used in space heating to recover heat generated during the manufacture of products for use in conditioning air. Energy does not typically change phase in a waste heat recovery system (air-to-air, water-to-water).

Five common types of heat exchangers are used in waste heat recovery systems:

1 Heat pipe exchangers: Heat pipe exchangers are high-efficiency thermal conductors, most commonly found in the form of evacuated tube collectors. The heat pipe is mainly used in space, process, or air heating.
2 Recuperators: Recuperators consist of a variety of different types of heat exchangers consisting of metal tubes that preheat the inlet gas before it is used in an industrial process or operation.
3 Economizers: An economizer is a type of heat exchanger where the waste heat in the exhaust gas is passed along a recuperator, reducing the energy required to heat the intake fluid for boilers or other equipment.
4 Regenerators: A regenerator is a type of heat exchanger that reuses the same stream of exhaust air after processing. In this type of heat recovery, the heat is regenerated and reused in the process.
5 Heat pumps: Heat pumps use an organic fluid that boils at a low temperature, allowing heat energy to be recovered from waste fluids.

Advantages of waste heat recovery are as follows:

■ Reduces the amount of flue emissions
■ Reduces equipment size for applications using the recovered heat
■ Reduces overall energy consumption and fuel cost

Tobacco Smoke Control

Secondhand smoke from cigarettes and cigars has largely been banned from the workplace and public buildings. Where it still exists, tobacco smoke represents a significant health hazard for employees and an equally significant liability hazard for employers. Even where smokers are only allowed to smoke in dedicated indoor rooms or outdoors only, businesses should follow smoke control procedures recommended by the USGBC and other organizations. Following are recommended environmental tobacco smoke (ETS) control techniques (adapted from the USGBC LEED new construction requirements)[10]:

- Prohibit smoking in the building completely.
- Locate any exterior designated smoking areas at least 25 feet away from any building entry, outdoor air intakes, or operable windows.
- Prohibit outdoor smoking except in designated smoking areas (necessary to prevent smoke from being pulled into the building through air intakes or windows).
- Locate designated smoking rooms to effectively contain, capture, and remove ETS from the building.
- Design smoking rooms to exhaust directly to the outdoors with no recirculation of ETS-containing air to the nonsmoking area of the building.
- Enclose the smoking room with deck-to-deck partitions that are sealed to prevent smoke escape from the ETS room.
- With the doors to the smoking room shut, operate an exhaust system sufficient to create a negative pressure (with respect to the adjacent spaces) of at least an average of 5 Pa (0.02 inch of water gauge, or 0.051 centimeter of water gauge), and with a minimum of 1 Pa (0.004 inch of water gauge, or 0.010 centimeter of water gauge).

PCB Removal

Polychlorinated biphenyls (PCBs) are a collection of 209 similar synthetic organic chemical compounds (known as congeners). Exposure to any of these compounds can cause harmful effects and varying levels of risk for humans. PCBs can alter a variety of body organs and functions, including the immune, hormone, nervous, and enzyme systems. Specific PCB health effects include multiple cancer types, Parkinson's disease, liver damage, diabetes, immune system damage, neurological problems in children, endometriosis, hearing damage, and heart disease. See Table 2.10 for tips relating to PCB hazards.

Tip Box

TABLE 2.10   PCB WORKPLACE HAZARD TIPS
■ Fluorescent light fixture ballasts (pre-1979)
■ Oil capacitors in various equipment and electrical devices
■ Well submersible pumps
■ Major electrical capacitors and transformers
■ Old industrial sites and building walls
■ Demolition waste and old building products

There are no known natural sources of PCBs. They were used in a wide range of industrial materials for over half a century. Materials that once held PCBs included: cutting oils, sealants, inks, paints, pigments, hydraulic fluids, and dielectric fluids used in electrical equipment. Because of their resistance to thermal breakdown and their insulating properties, they were the first choice for fluids used in transformers and capacitors. Because of their fire-resistance properties, PCBs were even required by fire codes in some areas. Concerns over the health effects led to a ban on their use in North America in 1977. By the mid-1980s, an initiative was underway to clean up contaminated areas and phase out PCB-containing equipment and products.

Although PCBs are no longer manufactured in the United States, people can still be exposed to them. The two main sources of exposure to PCBs are the environment and the workplace. Because of their resistance to degradation, PCBs can remain in the environment for decades. Past accidental discharges and careless disposal practices caused the release of PCBs into the groundwater and river and lake sediments. PCB contamination has been effectively treated by using systems incorporating activated carbon adsorption media. Activated carbon is widely used for the adsorption of many contaminants from liquids and air. Activated carbon is produced from carbonaceous organic substances, such as bituminous coal, bone, wood, coconut shell, or lignite. It is also used in many other applications, including food production and liquid decolorization.

Employers have an obligation to ensure that their workplaces are free of PCB contaminants. Fox River Watch, a project of the Wisconsin Clean Water Action Council, recommends a number of precautions for workers to avoid PCB exposure in their area. The ones specifically applicable to commercial facilities are as follows[11]:

CHECK OLD FLUORESCENT LIGHT FIXTURES

Most fluorescent light ballasts manufactured before 1979 contain roughly a teaspoon of concentrated PCBs sealed inside the capacitor. The capacitor is typically enclosed in a ballast box, which is surrounded by a tar-like material. Normal ballast operation does not emit measurable amounts of PCBs, but when an old ballast fails, the capacitor can rupture and leak PCBs. It may not be possible to tell the difference between tar and PCBs, so the safe course is to assume that any ballast leakage contains PCBs. The EPA banned the manufacture of equipment containing PCBs in 1979, so any ballasts manufactured after that date may be assumed to be free of PCBs.

CHECK EQUIPMENT CAPACITORS

PCBs may be present in oil-filled capacitors in old office, kitchen, and industrial equipment. Oil-filled (or running capacitors) are predominantly found in air conditioners, microwave ovens, dehumidifiers, submersible pumps, mercury vapor lamps, copy machines, and electrical control panels. Oil capacitors are rarely found in old refrigerators, dryers, washing machines, and fans.

CHECK SUBMERSIBLE WELL PUMPS

Submersible pumps with a two-wire design, manufactured before the 1979 ban, are likely to contain a PCB capacitor. When well pumps fail, the PCB capacitor can leak, contaminating the coolant inside the pump motor. If the motor seal fails, PCB-contaminated coolant can leak into the well water.

AVOID OR THOROUGHLY CHECK OUT OLD INDUSTRIAL SITES

Learn the history of industrial sites, old buildings, and property activities. Conduct environmental assessments of sites or site areas where work was conducted on old electrical equipment, capacitors, or transformers. PCBs can be absorbed several inches into building materials, such as old masonry, concrete, and tile, and they will volatize into the air for years.

AVOID ELECTRICAL EQUIPMENT CAPACITORS AND TRANSFORMERS

Old electrical transformers and capacitors may still, under statute, contain PCB oil for the life of the equipment (though new units may not). Avoid contact with any existing electrical equipment of this type, and have it replaced if possible and marked if it must remain in use. If PCB transformers or capacitors are involved in fires or explosions, assume contamination and clear the area until professional assessment and removal can be performed.

AVOID DEMOLITION WASTE EXPOSURE

Demolition waste, or old building products, may be PCB-contaminated. Painted products, old grout, old asphalt shingles, tar paper, pigments, and sealants may contain PCBs, occasionally at high levels. Separate stored demolition materials, ventilate the area, and have the materials tested for contamination. Never burn or melt unknown materials, since doing so may release airborne dioxins and furans into the air, both harmful PCBs.

Endnotes

1. Walker, Andy. Whole Building Design Guide: "Solar Water Heating." http://www.wbdg.org/resources/swheating.php. Accessed March 7, 2010.
2. Ibid.
3. Big Ass Fans. "Powerfoil X." http://www.bigassfans.com/powerfoil_x. Accessed April 28, 2010.
4. U.S. Environmental Protection Agency. "EPA's Asthma Program." http://www.epa.gov/asthma/programs.html. Accessed February 18, 2010.

5. U.S. Environmental Protection Agency. "Health Effects and Symptoms Associated with Mold Exposure." http://www.epa.gov/mold/append_b.html#Health Effects and Symptoms Associated with Mold Exposure. Accessed February 18, 2010.

6. U.S. Environmental Protection Agency. "Health Risks." http://www.epa.gov/radon/healthrisks.html. Accessed March 7, 2010.

7. Stanke, Dennis and Brenda Bradley. The Air Conditioning, Heating, & Refrigeration News, June 12, 2006: "Keeping Cool with Outdoor Air: Airside Economizers." http://www.achrnews.com/Articles/Technical/8feb3c885a7bc010VgnVCM100000f932a8c0. Accessed March 7, 2010.

8. Ibid.

9. North Carolina Department of Environment and Natural Resources. "Energy Efficiency in HVAC Systems." http://www.p2pays.org/ref/26/25985.pdf. Accessed February 18, 2010.

10. Siemens Corporation. "Environmental Tobacco Smoke (ETS) Control." http://www.us.sbt.siemens.com/siemensleed/06D90170.html. Accessed February 18, 2010.

11. Fox River Watch. "Reduce Your PCB Exposure and Protect Your Family" http://www.foxriverwatch.com/protection_pcbs_1.html. Accessed April 28, 2010.

3

UTILITIES

Utility costs are a primary factor in any commercial budget. Reducing those costs through efficiency, conservation, or on-site energy generation is the most useful form of sustainability any company can engage in.

See Table 3.1 for common forms of energy and their costs.

Electric Usage

Electricity is usually the greatest energy cost item for commercial and industrial businesses, as well as for agricultural operations. Whereas residential customers are only charged for electrical energy (measured in kilowatt hours, kWh), larger customers are also charged for demand (measured in kilowatts, kW). Demand is the instantaneous use of electricity, while energy is use over time. In Utah, for instance, electric rates for commercial and industrial customers of the Utah Power & Light Company are relatively high for demand but substantially lower for energy. Thus, a typical 80,000-square foot (7432-square meters) commercial office building may have a summertime monthly bill of $12,000, of which $8,000 is due to demand charges at $12.76 per kW and $4,000 due to energy use at $0.02574 per kWh. This corresponds to a composite equivalent rate of about $0.077 per kWh.[1]

A utility must be prepared to meet the highest demand on the grid, which usually happens on hot summer weekday afternoons when residential customers are air conditioning and commercial and industrial customers are still in full operation. To meet this demand, the utility must build power plants or buy power from other generators (which is quite costly during peak periods). Accordingly, rate structures are designed to charge commercial and industrial customers for their peak demand (typically averaged over a 15-minute period) over the past month.

Building managers in areas with overburdened electric systems have likely encountered the concept called demand response. Demand response (DR) is a set of activities tied to electric tariffs that seek to reduce electricity use outright, or at least shift heavy usage to less-intensive periods of the day. A local utility may provide DR control systems that encourage load shedding or load shifting during times when the electric grid

TABLE 3.1 COMMON FORMS OF ENERGY AND THEIR COSTS				
FUEL	UNIT	BTU/UNIT	COST/UNIT	COST/MBTU
Coal	Ton	28,000,000	$45.00	$1.61
Crude Oil	Barrel	6,300,000	$55.00	$8.73
Natural Gas	Therm	100,000	$1.00	$10.00
Heating Oil	Gallon	140,000	$1.80	$12.85
Propane	Gallon	92,000	$1.90	$20.65
Gasoline	Gallon	125,000	$2.25	$18.00
Electricity	kWh	3,412	$0.08	$22.56

is near its capacity or electricity prices are high. The overall intent of DR is to help businesses to manage their facility electric costs and to help utilities to improve electric grid reliability.

According to a 2007 study by the U.S. Demand Response Research Center, building heating, ventilating, and air conditioning (HVAC) operations consume one-third of the electricity used in most commercial facilities. HVAC therefore offers the most promising opportunity for DR energy reductions in a building. This is due to several factors, including the "flywheel effect," or the fact that shutting off a system does not have an immediate affect on the occupants or the energy management system in a facility.

Implementing DR strategies in office environments may mean adapting the HVAC control system to override its programmed goal of maintaining the design conditions at all times. Utilizing DR strategies in plant or process settings may require adjusting employee shift or production lines that have heavy electrical usage to times when the power is readily available at a lesser cost. Overall, global temperature adjustment of zones is the strategy that best achieves DR goals. In contrast, ad hoc adjustments to the air distribution or cooling systems can be helpful in DR, but building managers run a greater risk of causing discomfort to occupants. See Table 3.2 for a list of HVAC DR strategies.

Lighting represents roughly 40 percent of the energy consumption in the commercial building sector. The diversity of this sector presents some challenges to effectively mining energy-saving opportunities. For example, schools, hospitals, and office buildings have varying lighting requirements based upon the workspaces in question. Reducing lighting loads also carries the dual benefit of reducing heating and building cooling loads as well. A study by the U.S. Lawrence Berkeley National Laboratory estimated that, on a national annual average, each 1-kWh lighting savings produced a corresponding 0.48-kWh savings in cooling in a commercial building.[2]

In addition, design and efficiency considerations regarding a new construction versus retrofit situation differ greatly based upon the fixture layout scheme. As a result, national, state, and regional programs encourage the use of energy-efficient lighting technologies through a multitude of program approaches and measures to

Tip Box

TABLE 3.2   HVAC DEMAND REDUCTION STRATEGIES

CATEGORY	DR STRATEGY	DEFINITION	A	B	C	D
Zone control	Global temperature adjustment	Increase zone temperature setpoints for an entire facility.	X	X	X	X
	Passive thermal mass storage	Decrease zone temperature setpoints prior to DR operation to store cooling energy in the building mass, and increase zone setpoints to unload fan and cooling system during DR.	X	X	X	
Air distribution	Duct static pressure decrease	Decrease duct static pressure setpoints to reduce fan power.		X		X
	Fan variable frequency drive limit	Limit or decrease fan variable frequency drive speeds or inlet guide vane positions to reduce fan power.	X		X	
	Supply air temperature increase	Increase supply air temperature setpoints to reduce cooling load.	X	X	X	
	Fan quantity reduction	Shut off some of multiple fans or package units to reduce fan and cooling loads.	X	X	X	
	Cooling valve limit	Limit or reduce cooling valve positions to reduce cooling loads.	X			
Central plant	Chilled water temperature increase	Increase chilled water temperature to improve chiller efficiency and reduce cooling load.	X	X		
	Chiller demand limit	Limit or reduce chiller demand or capacity.	X			
	Chiller quantity reduction	Shut off some of multiple chiller units.	X	*	*	
Rebound avoidance	Slow recovery	Slowly restore HVAC control parameters modified by DR strategies.	**	**	**	**
	Sequential equipment recovery	Restore HVAC control to equipment sequentially within a certain time interval.	**	**	**	
	Extended DR control period	Extend DR control period until after the occupancy period.	**	**	**	

* The strategy can be applied to package systems by shutting off fewer compressors.

** Applicability of rebound avoidance strategies is determined by the DR strategies selected.

HVAC SYSTEM TYPES

Type A	Constant air volume system with central plant
Type B	Variable air volume system with central plant
Type C	Constant air volume system with package units
Type D	Variable air volume system with package units

attain energy savings. The U.S. Department of Energy participates in operating an online directory of national and state energy incentives and rebates. Information is available at www.dsireusa.org/.

Natural Gas Usage

The U.S. Department of Energy's Federal Energy Management Program offers the following general tips to reduce facility natural gas consumption[3]:

BUILDING EFFICIENCY

Overall building efficiency is sometimes ignored by building managers in favor of more specific and occasionally, exotic, strategies in pursuit of easily quantifiable savings. Features such as photovoltaic solar panels are certainly a more visible sign of an intent to save energy, but simple energy conservation measures offer far more bank for the money. Following are a handful of simple economy measures every commercial facility should employ to minimize their energy consumption and operational costs. See Fig. 3.1 for a summary of these techniques.

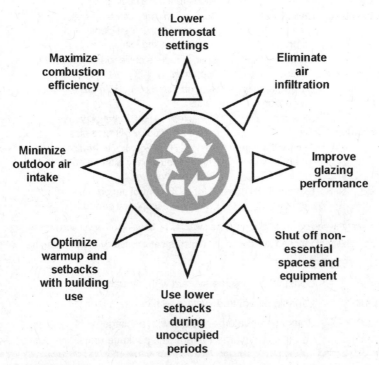

Figure 3.1   Building energy-efficiency tips.

Lower thermostat settings   In consultation with building users, adjust thermostats for lower temperatures in the winter. For each degree reduction in the thermostat setting, a 3 percent reduction in fuel consumption can be achieved.

Lower setback temperatures during unoccupied periods   For a typical building, a 10 percent reduction in annual fuel consumption can be achieved if the thermostat setting is lowered 10°F (4.45°C) for an average of eight hours each day.

Maximize combustion efficiency   Annually inspect and perform preventative maintenance on furnaces, water heaters, and space heaters. Building managers should change filters every six months. Preventative maintenance should include filter changes, as well as adjusting natural gas burners for proper excess air settings and efficient combustion. The potential gas savings from a diligent preventative maintenance program: 2 to 12 percent of annual fuel use.

Reduce glazing solar gain and conductive loss   Use exterior shading devices, interior blinds, reflective glazing film, or other techniques to both reduce solar heat gain during the summer and conductive heat loss through the glass during the winter. If the existing building glazing is inferior (single glazing without thermal breaks), consider upgrading or installing storm sashes or additional lites to improve performance.

Optimize morning warm-up and night setback controls   Ensure that the programmable temperature controls, energy management settings, and control systems for the facility are synced with the production schedules and actual building usage. This will ensure that the heating system is not operating when the building is not in use, or is not heating the building well before or beyond normal working hours. Adjust settings for daylight savings, time changes, and when plant vacations or other shutdowns occur. On a more detailed level, fuel savings can also be achieved by updating the warm-up and setback control schedules to coincide with current occupancy periods in affected buildings for each heating zone and weekday.

Minimize outdoor air use for ventilation   Installations that use all or a large percentage of outside air to ventilate hazardous areas may be wasting a substantial amount of gas in excess heating. A close examination of the ventilation systems may reveal the possibility of recirculating and filtering some exhaust air within the facility.

Shut off nonessential equipment and spaces   Isolate unused building areas by reducing space temperature, providing only minimum heat levels to protect against freezing.

Eliminate major sources of infiltration   Leakage of outside air into heated spaces on cold days can be a substantial heating load contributor in some buildings. Annually inspect and maintain door, window, and building joint sealants. Keep large overhead doors in warehouses and industrial buildings tightly closed. Install air curtains or wind barriers at building openings where a door is impractical for efficient operation. Periodically check and repair overhead door seals.

CENTRAL HEATING PLANTS

Optimize combustion efficiency It is important to maintain minimum and constant, excess air levels in boiler and furnace systems to assure complete combustion. However, allowing too much air into the system will result in excess gas consumption. This is a common problem. With well-designed natural gas-fired boilers, an excess air level of 10 percent is usually considered optimal. Excess air levels should be continuously monitored by maintenance personnel.

Minimize boiler blowdown Reliable steam plant operation requires that a portion of the boiler water be drained periodically to minimize the concentration of solids. Excessive blowdowns waste fuel and create more plant downtime. Plant personnel should develop a minimum blowdown schedule to ensure that excess fuel consumption for this purpose is minimized.

Optimize boiler loading Selected boilers should be shut down during periods of low heating loads. This allows the remaining boilers to operate more efficiently at higher rates.

Perform boiler maintenance The combustion chamber and heat transfer surfaces should be cleaned at regular intervals. Stack temperatures of more than 150°F (65.6°C) above steam temperature may indicate the presence of excessive water-side scaling. Scaling can reduce heat transfer and increase fuel consumption by as much as 10 percent.

Inspect/replace steam traps Steam traps are mechanical devices that remove condensate from steam piping and equipment. A typical system may include hundreds of steam traps, and as many as 15–20 percent of these devices may not operate properly. Collectively, steam trap energy losses result in a significant excess energy cost for a facility. A scheduled preventative maintenance program can reduce the number of nonworking steam traps to no more than 5 percent of the total.

Repair condensate return equipment Inoperative condensate return equipment, like steam traps, often goes unnoticed because collected condensate flows directly into the drain system largely unnoticed. Condensate contains valuable heat energy that can be recovered to offset process fuel costs. If a facility installs a system to capture condensate and return it to the steam plant, fuel costs can be reduced by as much as 10 percent.

Repair steam leaks   Steam leaks can be a significant source of efficiency loss and excess gas consumption, as well as posing an employee safety hazard.

Repair insulation Up to 25 percent of total heating system fuel costs can be blamed on thermal losses at distribution piping, valves, and equipment. Thermographic instruments and infrared pyrometers can be helpful in surveying steam lines and identifying areas where insulation can be installed or repaired to reduce these losses.

Isolate nonessential distribution piping   Process and production changes often result in changes to the steam piping over time. Such alterations may result in overall systems that are no longer optimally designed for the total load, wasting energy and resulting in steam runs that are unnecessarily long or with more bends than are required. Plants may find that opportunities are available to eliminate major sections of a distribution system originally designed to supply much larger loads, allowing existing loads to be served in a more efficient manner. Distribution losses from inefficient systems can be eliminated through a periodic reassessment of the overall layout and repiping for more efficient runs.

Reduce distribution pressure   Load reductions for changing processes and equipment eliminations may also offer the opportunity to reduce steam pressures in existing distribution systems to achieve a reduction in thermal losses.

Recommended gas equipment efficiency ratings
- Combined High-Efficiency Boiler and Water Heating Unit with an Annual Flue Utilization Efficiency (AFUE) rating greater than or equal to 90 percent.
- Combined High-Efficiency Boiler and Water Heating Unit with an AFUE rating greater than or equal to 85 percent.
- Natural Gas Hot Water Boiler with an AFUE rating greater than or equal to 90 percent.
- Natural Gas Warm Air Furnace with Electronic Commutated Motor (ECM) and an AFUE rating greater than or equal to 94 percent.
- Natural Gas Hot Water Boiler with an AFUE rating greater than or equal to 85 percent.
- Condensing Natural Gas Water Heater with a thermal efficiency of 95 percent or greater.
- Natural Gas Warm Air Furnace with ECM and an AFUE rating greater than or equal to 92 percent.

See Table 3.3 for an estimation guide on savings that can be achieved through upgrading system AFUE ratings.

Improve furnace efficiency   Seco/Warwick of Meadville, Pennsylvania, offers the following tips for improving efficiency in aluminum furnaces, though the advice is applicable to many industrial furnace applications[4]:

Gas consumption reduction tips   Building managers should track historical gas consumption and periodically compare the records to current consumption as a way to monitor efficiency. The British Thermal Units (BTUs) required to operate the furnace will vary with the load size in the furnace, the type of furnace, and the temperature setting. Each application is unique, so maintaining usage records over time is the best way to monitor efficiency for a particular installation. Seco/Warwick notes the following as

ROI Quik-Calc

TABLE 3.3	SAVINGS RESULTING FROM AN AFUE SYSTEM UPGRADE			
	PROPOSED AFUE			
	80%	85%	90%	95%
50%	$35	$40	$43	$45
55%	$30	$35	$38	$40
60%	$25	$30	$33	$35
Existing 65%	$20	$25	$27	$30
AFUE 70%	$10	$15	$20	$25
75%	$5	$10	$15	$20
80%		$5	$10	$15
85%			$5	$10

Divide total existing equipment utility cost by 100. Multiply the savings in each box by this value to estimate total overall savings from AFUE upgrade.

general guidelines for atmosphere consumption in efficient furnaces. A furnace requiring more atmosphere should be checked for leaks or a badly adjusted flow control system:

- Furnace purging: 4-5 furnace volumes
- Temperature maintenance during heating: $1/2$ to 1 furnace volumes
- Cooling: $1 1/2$ to 2 furnace volumes

Gas consumption   An increase in gas consumption may result from the following causes:

- **Lean combustion system.** Lean combustion systems may have too much excess air, requiring more BTU consumption to heat the air to the process temperature. Most combustion systems can run with as little as 10 percent excess air.
- **Excess positive furnace pressure.** Furnace pressure should be adjusted to the minimum level necessary for the process. Higher than required furnace pressure causes heated furnace atmosphere to escape around furnace openings, reducing the efficiency.
- **Negative furnace pressure.** Conversely, negative furnace pressure pulls ambient air into the furnace, forcing the burners to consume more gas to maintain temperature. It is important to monitor pressure dampers or electronic pressure controls to maintain a slight positive furnace pressure.
- **Load sizes.** More fuel is required to process a small load in a large furnace, so industry should maximize load sizes as much as possible.
- **Soak period.** BTU consumption per pound goes up if there is a long soak period.
- **Furnace air leaks.** Check for leaks at door seals, gaskets at plugs and access plates, radiant tubes, welded joints, shaft seals at fans, dampers, and unwelded connections.

Water Usage

Reducing water consumption for industrial processes is a goal that requires close coordination between industrial/process engineers and managers. Almost every production facility uses process water to some degree to either clean product or enable their process to run more efficiently. While more efficient piping and equipment is certainly an option that businesses should examine in reducing plant water consumption, the most substantial benefits can often be achieved through examining ways to simply reduce water consumption or use process water for more than one application. Following are some examples of water savings for common types of industrial applications.[5]

OFFICE, RETAIL, AND INSTITUTIONAL BUILDINGS

Toilet, kitchen, and landscape uses are predominant in office building, retail, and institutional occupancies. In residential living facilities such as hotel/motel, assisted living, or apartments, showers, baths, and dishwashing (hot water consumption) become primary uses. Consider these steps to reduce consumption in these occupancies:

- Low-flow faucets and shower restrictors
- Waterless toilets (See Fig. 3.2 for a photograph of a waterless urinal)

Lower hot water heater setting

- Insulate hot water tank and lines
- Convert to tankless (instant-supply) hot water heater
- Convert to ENERGY STAR dishwasher and washing machines
- Employ Xeriscaping (or low-water-usage) plantings
- Discontinue using site irrigation systems

Legionnaires' disease (Legionellosis)   A common energy-saving strategy in commercial buildings is to lower the hot water temperature for either potable or heating systems. At lower settings, bacteria may grow and cause serious illness among the building occupants. The *Legionella* bacteria are found naturally in the environment and grow best in warm water typically found in cooling towers, hot water tanks, extensive plumbing systems, or ductwork of the air-conditioning systems of large buildings.

The first known outbreak of Legionnaires' disease (now called Legionellosis) occurred in July 1976 at an American Legion meeting at the Bellevue Stratford Hotel in Philadelphia, Pennsylvania. This initial outbreak killed 34 people and sickened over 220. Fatality ranges for outbreaks of Legionellosis range from 5 to 30 percent, making it a potentially lethal hazard in commercial environments. Building users get Legionnaires' disease when they breathe in mist or vapor droplets that have been contaminated with the bacteria, making building HVAC systems a ready source of spreading for the bacteria. The range of water temperatures in which *Legionella* cannot survive is 158 to 176°F (70 to 80°C).

Figure 3.2 Waterless urinal.

METAL FINISHING

The metal-finishing industry offers many opportunities to reduce both water-use and pollution-abatement costs. The plating and anodizing processes involve a multitank, multistep process. Water-saving industrial practices include:

- Drag-out control
- Good tank design
- Efficient rinse practices

- Process controls and meters
- Chemical recovery
- Good exhaust-hood design

The major reason to dump water is contamination of a rinse or process tank with liquid from the previous tank. Methods include:

- Designing racks, baskets, and barrels so parts drain and do not retain liquids
- Using turning, tilting, and "bumping" to remove excess liquid
- Using drip or drain boards to collect and drain liquids back into the source tank
- Allowing parts to remain over the tank for a few seconds (dwell time)
- Washing or blowing contaminants back into process or dead tanks using fogs, sprays, or air knives
- Using wetting agents
- Using chemicals or heat to reduce plating-solution viscosity
- Operating the solutions at minimum possible concentration

The following tank-design methods reduce water use or allow for better reuse and recovery of metals:

- Using air or mechanical agitation to promote mixing and good contact
- Hard plumbing all piping so hoses cannot be left on inadvertently
- Preventing short-circuiting of fluids
- Sizing tanks to the minimum for the pieces to be plated
- Segregating waste streams so both metals and water can be recovered more easily

Efficient rinsing saves water and chemicals and reduces wastewater costs. Methods involve several technologies, including:

- Using sprays on flat pieces of metal
- Using counter-current rinsing, where the piece is rinsed in successively less concentrated tanks, with the water from the first tank being used as feed for the second, and so on
- Using reactive rinsing, where the rinse water from the final tank is used for the pickle-rinse tank and the pickle-rinse tank water is used as feed to make up the alkaline-rinse tank
- Using air agitation of the tanks

Flow- and process-control opportunities include:

- Installing conductivity controllers to discharge water only if the chemicals have become too concentrated
- Metering makeup water for good process control and to identify problems
- Using flow restrictors to limit the amount of water being added

Chemical and water recovery includes several water-treatment technologies:

- Filtering plating fluid to remove suspended matter
- Using membrane technology to recover metals and water
- Using reverse osmosis (RO) or deionization for the feed water for both rinse and process-fluid tanks to reduce interference from other ions
- Regenerating spent acids
- Using RO reject water for cleanup around, but not in, the tanks

Scrubber system design can also reduce water use by:

- Recirculating scrubber liquid
- Using scrubber water above plating tanks as makeup water for that process
- Using RO reject water or similar reject streams as makeup water for scrubbers for which scrubber effluent will not be reused

PULP AND PAPER MANUFACTURING

Producing paper from pulpwood and other fiber sources is the beginning process for all paper. It is also the most water- and energy-intensive stage in the life of a paper product. Five gallons of water are used to make one pound of paper (according to Weyerhaeuser Corporation). Recycling paper and cardboard products cuts this energy and water use in half. According to Conservatree, an organization that promotes paper recycling, most repulping for recycled paper is done at pulp and paper mills where paper is made, or in special facilities that use the product to make such things as cellulose insulation or .pulp products, such as egg cartons.

The practices discussed above reduce water use by decreasing the amount of cleaning required at the end of the press run. Printing operations also produce large amounts of waste heat in cooling the equipment. Large operations often have cooling towers.

FOOD AND BEVERAGE PROCESSING

The food-and-beverage industry uses water for many purposes. The quality and purity of the water is of primary concern since it is used to make products that will be consumed. Water is also used to clean and sanitize floors, processing equipment, containers, vessels, and raw food products prior to their processing.

Hot water, steam, cooling, and refrigeration systems all require source water. Designing and building a facility that has a reduced requirement for water includes:

- Incorporating water reuse and recycling
- Taking advantage of dry methods for cleanup and transport
- Designing the facility for ease of cleaning

■ Providing adequate metering, submetering, and process controls
■ Using product and by-product recovery systems

Because of the complicated and highly varied nature of the food-and-beverage manufacturing industry, providing a simple guide to water efficiency that covers all types of facilities is not possible. Before beginning this discussion of water conservation in food processing, one should remember that health and sanitation are overriding concerns. All actions to reduce water use must be measured against this primary consideration.

The following example illustrates ways water can be used in the soft drink industry. Potable water is first treated to soften it and, if needed, to remove additional minerals. It is chilled and blended with flavorings and sweeteners before being carbonated. Cans or bottles are filled and sealed, then rinsed and sent through a warming bath to avoid creating condensate in the open air and ensure they are dry before packing. The eight major water-using processes in the soft drink industry are:

■ Water softening, which requires periodic filter backwash
■ Water included in the product
■ Water to clean and rinse cans
■ Water to warm cans after processing
■ Water sprayed on the conveyor line as a lubricant
■ Water to operate cooling towers for refrigeration equipment and boilers for heat
■ Water to sanitize and clean the plant and vessels
■ Water for employee sanitation and irrigation

The use of flumes both to transport and clean produce (fruits and vegetables) is common. Water is also used in the cleaning and processing of meat, poultry, and fish. Common water-conservation techniques begin with reducing water use by:

■ Recycling transport water
■ Adjusting design of flumes to minimize water use
■ Using flumes with parabolic cross-sections
■ Providing surge tanks to avoid water loss
■ Using float control valves on makeup lines
■ Using solenoid valves to shut off water when equipment stops

All these techniques can reduce the need for water, but changing the process has even more potential.

■ Control process equipment to reduce waste.
■ Replace fluming with conveyor belts, pneumatic systems, or other dry techniques to move food products.
■ Control sprays on belts.
■ Install food washing sprayers.
■ Use mechanical disks and brushes to reduce consumption.
■ Install counterflow washing systems.

Equipment cleaning   Where water is used for cleaning, it is important to employ the "multiple aliquots" concept, in which it is better to use a number of smaller volumes of water to clean than one very large volume. For mixers, extrusion and molding equipment, conveyor belts, and other open equipment to which one can gain direct access, cleaning should start with physical removal of residual materials and then be followed by wet washing. Four principles of wet cleaning are:

- Use high-pressure, low-volume sprays.
- Install shutoffs on all cleaning equipment.
- Use detergents and sanitizing chemicals that are easily removed with minimum water.
- Install and locate drains and sumps so water and wastes enter quickly to prevent the use of a hose as a broom.

Container (bottles, cans, cartons) cleaning   Cleaning bottles, cans, and containers prior to filling them is common throughout the industry. For returnable bottles, the use of air bursts to remove loose debris and materials and the reuse of water from can warming and other operations are common ways to reduce water use. Other methods include use of pressure sprays and steam instead of high volumes of hot water to clean containers.

Can and bottle warming and cooling   Water has a variety of applications, ranging from cooling or heating cans to use as a heat-transfer agent. This water remains relatively clean and is an excellent source of water for reuse. Water is used to cool cans after they have been removed from pressure cookers in the canning process. In most cases this water is cooled in a cooling tower or a refrigeration unit that employs a cooling tower in the process. In the warming process, cans and bottles from the beverage industry that have been filled with cold liquids are heated so condensate does not form on them and they dry more quickly before packing. These operations offer significant opportunities for reuse for almost all of the other water needs in the operation, except where potable quality is required by regulation. Examples of reuse include:

- First rinse in the wash cycle
- Can and bottle shredder and crusher operations
- Filter backwash for product filters
- Chemical-mixing water
- Defrosting of refrigeration coils
- Use for equipment or floor cleaning
- Flushing out shipment containers and crates
- Cleaning of transport truck and rail cars
- Gutter and sewer flushing
- Fluming and washing of fruits and vegetables
- Makeup water for conveyor lubrication systems
- Irrigation
- Cooling-tower makeup water

Water-savings potential summary   Examples of practices and water savings are provided above. Because of the varied nature of the products and processes found in the food-and-beverage-processing industry, water-savings potential is slightly different for each. These five design principles will help build water efficiency into a facility:

■ Design the facility for ease of cleaning.
■ Provide adequate metering, submetering, and process control.
■ Set up the facility to take advantage of dry methods for cleanup and transport.
■ Use product and by-product recovery systems.
■ Incorporate water reuse and recycling.

AUTOMOTIVE SERVICES

Three areas of operation offer both reduced water- and pollution-loading possibilities:

■ Proper design of aqueous parts- and brake-cleaning
■ Preventing pollution and reducing water use in shop-floor cleaning
■ Proper handling of spent fluids and oils

Keeping floors clean in the first place eliminates the need for frequent washing. Methods include:

■ Installing secondary containers under fluids-storage containers to catch leaks and using drip plans under vehicles being worked on
■ Using dry cleanup with hydrophobic mops for oil and using absorbent materials (kitty litter, rice hulls, pads, rags, pillows, and mats) to cleanup spills
■ Sealing floors with an epoxy material, which significantly aids in cleanup and prevents oils and liquids from penetrating concrete floors
■ Providing floor-cleaning equipment that scrubs and vacuums up its own water
■ Eliminating the use of open hoses for cleanup and using pressure-washing equipment infrequently and for major cleanup events only
■ Marking drains clearly to ensure that floor drains are differentiated from storm drains and all floor drains are connected to an oil separator

Recovery and recycling of radiator flush-water both saves water and reduces pollution-loading. Using storage vessels designed to hold spent antifreeze and other fluids, such as oil and transmission fluid, both eliminates the need to clean and flush these fluids down a drain and is required as part of modern pollution-control methods. Water use in facilities that recycle radiator flush-water has been shown to be less than 10 percent of water use in nonrecycling facilities.

Sustainability best practices
■ Require new facilities to provide secondary containers to catch drips, leaks, and spills from stored liquids and solvents.
■ Seal shop floors with acrylic sealers for easier cleaning.

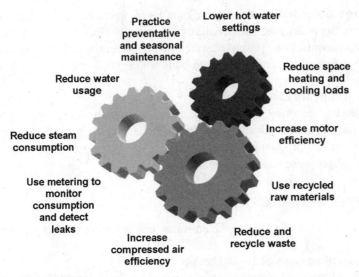

Practice preventative and seasonal maintenance

Lower hot water settings

Reduce space heating and cooling loads

Reduce water usage

Increase motor efficiency

Reduce steam consumption

Use metering to monitor consumption and detect leaks

Use recycled raw materials

Increase compressed air efficiency

Reduce and recycle waste

Figure 3.3   **Generic manufacturing facility sustainability techniques.**

- Require automatic shutoff and solenoid valves on all hoses and water-using equipment, where applicable.
- Require aqueous parts- and brake-cleaning equipment to employ recirculating filtration to minimize the need to dump water.
- Properly identify all drains and cleanouts.
- Have proper facilities for the capture, storage, and recycling of spent fluids, oils, and fuels, including antifreeze and radiator flush-water.

Many other industries can benefit from introducing energy efficiency techniques and sustainability practices. See Fig. 3.3 for a list of generic practices that can be implemented at a wide range of manufacturing facilities.

Renewable Energy

Commercial facilities account for approximately 72 percent of the electric consumption in the United States. Energy policies at the state and federal levels have demonstrated one overriding trend—promoting the increased use of renewable energy.

As of 2007, renewable energy sources provide about 6.9 quadrillion BTUs, or about 7 percent of the U.S. total energy need. The following types of renewable energy comprise this total[6]:

- Biomass: 53 percent
- Hydroelectric: 36 percent
- Wind: 5 percent

■ Geothermal: 5 percent
■ Solar: 1 percent

As of mid-2009, more than 29 states had enacted mandatory renewable energy port-folio standards for utilities.[7] In 2002, for instance, California enacted a requirement that state utilities produce 20 percent of their power from renewable sources by the year 2017. California already leads the nation in the development of renewable energy. As of 2003, 27 percent of the state's total energy consumption was produced by renewable sources (including hydropower). The increasing trend toward requiring utilities to offer renewable energy options to businesses and consumers is likely to increase further, especially with enactment of the renewable energy provisions of the American Recovery and Reinvestment Act of 2009. The provisions were:

1 Extension of the wind energy production tax credit (PTC) to 2012 and the PTC for municipal solid waste, qualified hydropower, biomass, and geothermal energy to 2013. The wind PTC had been set to expire by the end of 2009.
2 A two-year extension of the PTC for marine and hydrokinetic renewable energy systems through 2013.
3 The Act allows owners of nonsolar renewable energy facilities to make an irrevo-cable decision to earn a 30 percent investment credit rather than the PTC. The option remains in effect for the current period of the PTC.

Business managers may find U.S. state and federal tax advantages to using renew-able energy sources at their facilities. Developing on-site renewable energy sources is certainly one option (see Chap. 2), but with the growing offerings of renewable energy portfolios from utilities, businesses may find purchasing renewable energy directly through the grid to be the most advantageous course. Renewable power is typically sold at a premium compared to nonrenewable power sources (though in some areas power generated by hydroelectric sources is competitive with coal-fired electric generation). This premium may be completely offset, however, by tax ben-efits offered through state or federal governments. Existing facilities can earn 1–4 points under LEED Existing Building certification program by installing on-site renewable energy sources. See Table 3.4 for a summary of these requirements.

RENEWABLE ENERGY CREDITS

Although energy policy in the United States is very changeable, and currently a subject of much political debate, some trends favorable to business use of renewable energy appear unlikely to change.

A renewable energy credit (REC), sometimes referred to as a renewable energy certificate (or greentag), is an environmental commodity that represents the added value, environmental benefits, and cost of renewable energy above conventional methods of producing electricity, namely, burning coal and natural gas. RECs help wind farms and other renewable energy facilities grow by making them more financially viable, thereby incentivizing development.

A REC represents the property rights to the environmental, social, and other non-power qualities of renewable electricity generation. A REC, and its associated attributes

TABLE 3.4   LEED EXISTING BUILDING RENEWABLE ENERGY CREDITS

- Meet some of the building's total energy needs with on- or off-site renewable energy.
- Off-site renewable energy must be Green e-certified by the Center for Resource Solutions.
- The environmental attributes of on-site renewable energy must be retained. Credits cannot be sold.
- Up to four LEED points are available:
  - One point for 3% on-site or 25% off-site renewable energy
  - Two points for 6% on-site or 50% off-site renewable energy
  - Three points for 9% on-site or 75% off-site renewable energy
  - Four points for 12% on-site or 100% off-site renewable energy

and benefits, can be sold separately from the underlying physical electricity associated with a renewable-based generation source.

RECs provide buyers flexibility in:

- Procuring green power across a diverse geographical area.
- Applying the renewable attributes to the electricity use at a facility of choice.

This flexibility allows organizations to support renewable energy development and protect the environment when green power products are not locally available. As renewable generators produce electricity, they create one REC for every 1000 kWh (or 1 megawatt-hour) of electricity placed on the grid. If the physical electricity and the associated RECs are sold to separate buyers, the electricity is no longer considered "renewable." The REC product is what conveys the attributes of the renewable electricity, not the electricity itself.

RECs and the attributes they represent are an ingredient of all green power products. REC providers—including utilities, REC marketers, and other third-party entities—may sell RECs alone or sell them bundled with electricity. As of 2007, more than 50 percent of utility customers had access to green power bundled products. All customers had access to buying RECs.

Buyers can identify green power suppliers using U.S. Environmental Protection Agency's (EPA's) Green Power Locator tool available at www.epa.gov/greenpower/pubs/gplocator.htm.

There are a number of third-party organizations in the market that certify RECs. As a best practice, EPA recommends that buyers seek out certified products as a form of buyer protection. Certified RECs should meet national standards for resource content and environmental impact. Certification ensures that RECs meet accepted standards of quality.

GREEN POWER VERSUS RENEWABLE ENERGY

Renewable electricity is produced from resources that do not deplete when their energy is harnessed, such as sunlight, wind, waves, water flow, and biological processes such as anaerobic digestion (e.g., landfill gas), and geothermal energy. Renewable electricity resources are distinct from fossil and nuclear fuels, which are also used to generate electricity.

EPA defines green power as a subset of renewable electricity and represents those renewable resources and technologies that provide the highest environmental benefit. Green power is renewable electricity produced from solar, wind, geothermal, biogas, biomass, and low-impact small hydroelectric resources. Definitions for renewable energy can vary and may include resources that are acknowledged to have some environmental impacts, such as land use and fisheries impacts of large hydro dams.

Renewable energy facilities generate renewable energy credits or certificates (RECs) when they produce electricity. Purchasing these credits is the widely accepted way to reduce the environmental footprint of your electricity consumption and help fund renewable energy development. Purchasing RECs at the same quantity as a facility's electric consumption guarantees that the energy used is added to the power grid from a renewable energy facility, and supports the further development of these facilities.

On-site renewable energy electric production for commercial or industrial facilities is usually tied into the grid to best realize cost savings for the company. Greater savings, however, can usually be had through direct use of the generated power in those instances where on-site uses can be linked to both renewable and backup power, or energy consumption at the facility is low enough where battery storage can accommodate periods when the renewable system is not generating sufficient power. See Table 3.5 for a summary of situations where facilities of various types may be able to use direct energy generated from an on-site system.

Tip Box

TABLE 3.5   FACILITY RENEWABLE ENERGY SOURCES

	HYDRO	WIND	PV	GEO-THERMAL	BIOMASS	SOLAR THERMAL
Mini-grid power for commercial, industry, process	•	•	•	•	•	•
Process heat, cogeneration				•	•	•
Unattended loads	•	•	•	•		•
Individual systems for residential, small commercial, isolated plant systems	•	•	•		•	
Water pumping, water treatment	•	•	•		•	•
Space heating, water heating	•	•		•	•	•

Whether purchased from Renewable Choice or a local utility's "Green Pricing" program, both systems are based on the purchase of RECs. Renewable Choice offers the same product as a local utility company but at a fraction of the price and with numerous value-added services.[8]

On-Site Energy Generation

See Table 3.6 for an *ROI Quik-Calc* of on-site energy generation potential.

WIND ENERGY

Wind energy can be captured on-site using wind turbines erected on tall towers to capture prevailing winds. Wind energy is cost-effective in areas with adequate wind resources. The U.S. Environmental Protection Agency (EPA) estimates that a 3-kW turbine with a 60- to 80-foot tower could reduce a facility's monthly electricity bill by 30 to 60 percent (assuming monthly electricity costs range between $60 and $100). A system of this capacity could generate approximately 700 to 1100 kWh. Using 2006 values for the national average installed cost for wind projects of approximately $1480 per kW capacity, the payback period for this system could be as short as six years.

Small wind turbines, those with a capacity of 100 kW or less, can be appropriate for single facilities as on-site energy generators. Although large-scale wind turbine installations may require a minimum of 1 acre of land and wind speeds averaging 49 feet per hour (15 meters per hour) at 164 feet (50 meters) above the ground, small turbines can be appropriately sited in urban areas and smaller sites, at lower heights, and with lower average wind speeds.

ROI Quik-Calc

TABLE 3.6   ON-SITE ENERGY GENERATION POTENTIAL
Electricity production and consumption (measured in kWh) are a function of generation capacity (measured in kW) and time (measured in hours). System generation capacity depends on a site-specific capacity factor, which describes the system's actual energy output divided by the output if the system operated at full capacity. Electricity production can be calculated as follows:
Electricity production (kWh) = Capacity (kW) × Capacity factor × Time (hours)
Solar photovoltaic panels typically have capacity factors between 0.07 and 0.17. Wind turbine capacity factors are between 0.25 and 0.30 (Hull 1 in Hull, Massachusetts, for example, operates at 0.27). Most fossil fuel power plants have capacity factors near 0.28.
Example: The annual electricity production of a 10-kW PV system with a capacity factor of 0.15 would be calculated as follows:
10 kW × 0.15 × 8760 hours = 13,140 kWh per year (36 kWh per day)
Source: U.S. Environmental Protection Agency: http://www.epa.gov/statelocalclimate/documents/pdf/7.2_on-site_generation.pdf

PHOTOVOLTAIC ENERGY

Photovoltaic (PV) systems convert sunlight directly into electricity in solar cells. PV systems can produce electricity even in weak sunlight, and the best of them can generate sizeable quantities of electricity under ideal conditions, including quality of the sunlight and the system's mounted pitch. The New York State Energy Research and Development Authority, for example, has estimated that PV systems in their state can produce between 1000 and 1300 kWh per kW capacity annually. Using this assumption, a 10-kW system capable of producing 1500 kWh per kW per year could produce 15,000 kWh annually. The EPA estimates that a 20,000-square foot (1858-square meter) office building using 15.5 kWh per square foot could reduce its purchased grid electric consumption by approximately 5 percent through using such a system.

PV systems are ideal for rooftop installations, lending them to ready use on industrial or other commercial facilities with large, flat roof areas. PV systems can be installed as stand-alone systems that are not connected to the electricity grid or as linked systems that resell power to the local utility when consumption is lower than supply. Small stand-alone systems may be suitable for supplying power for site lighting, irrigation, or other facility systems with limited power draws. See Fig. 3.4 for a photograph of a building that uses a PV array that doubles as window shading.

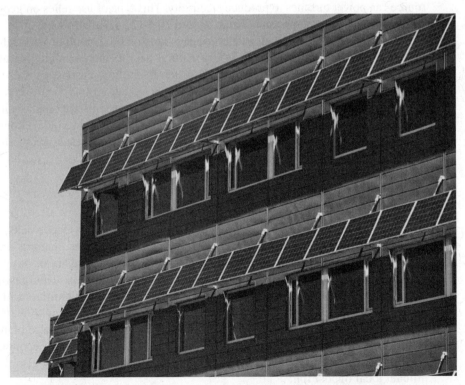

Figure 3.4   Building with a dual photovoltaic collector and shading device.

SOLAR HOT WATER

Passive solar hot water systems use the sun's energy to heat water in a collector. The heated water is distributed throughout the building directly, or used to charge a conventional hot water system to reduce the need for auxiliary heat to maintain the hot water system design temperature.

SOLAR HEATING AND COOLING

Solar process heating uses sunlight to heat a medium of contained air or fluid. The captured heat is distributed through a building via air or water systems, or used as a supplemental heat source to a conventional forced air or hot water system. PV systems may also be used to provide electricity to conventional electric-powered air conditioning systems.

GEOTHERMAL

Geothermal systems use the earth's stable below-ground temperatures in generating electricity, providing heat for input to building HVAC systems, and in heating hot water. The most direct use of geothermal energy (though less common outside of the western United States) is the use of steam from beneath the earth's surface. This direct source can power turbines to produce electricity. This type of use relies on geothermal reservoirs with water temperatures that exceed 70°F (21°C) at all times.

A second type of geothermal energy usage is much more common. This technology uses heat pumps to utilize below-ground natural heat to incrementally warm liquid to provide central space heating or to heat water. In cooling seasons, geothermal heat pumps can exchange warm surface air for cooler below-ground air to increase the efficiency of the cooling cycle. Geothermal heat pump systems are typically installed at shallow depths, between 4 and 6 feet (1.2 and 1.8 meters) below grade. Ground temperatures are fairly constant throughout the United States. As a result, geothermal heat pumps can be effective in almost every location.

BIOMASS

Electricity steam turbines can be fueled by burning solid biomass feedstocks, including construction wood waste, plant material, agricultural wastes, manure, and even sewage. Biomass generates electricity by heating the material in an oxygen-free environment to convert it into combustible oil or gas biofuels. This gasification process can be up to two times more efficient than burning solid biomass and results in reduced emissions. Although a promising technology in areas with a sufficient supply of feedstock, biomass operations are only appropriate for large plant sites, or areas where private/government partnerships make such projects economically feasible. Companies seeking to erect biomass-burning facilities may also face significant local opposition.

HYDROPOWER

Low-impact hydropower energy systems use the kinetic energy of moving water to produce electricity. Hydropower is renewable and produces few emissions, but the need to restrict water flow has environmental impacts on fish, water purity (buildup of nutrients), and changes in the local ecology. The Low-Impact Hydropower Institute (LIHI) provides certification to hydropower projects that successfully demonstrate minimal impact on the environment. The EPA Green Power Partnership recognizes hydroelectricity as green power when projects are LIHI-certified.

FUEL CELLS

Fuel cells combine oxygen and hydrogen to produce electricity without combustion. Fuel cells do require a continuous stream of hydrogen-rich fuel, and can only be considered a green energy technology if they operate using a hydrogen fuel generated by renewable sources, such as pure hydrogen generated by solar or wind energy generating systems or digester gas.[9]

Wastewater

The reduction of wastewater usage has become an increasingly more important component of site sustainability with the decrease in water availability across the United States (particularly in the western states), and the increasing cost of wastewater treatment that must be borne by businesses.

Sustainable wastewater practices include three basic components:

1 **Conservation:** Reduce the overall amount of water used by the facility, and therefore reduce the overall amount of wastewater produced.
2 **Reuse:** Reuse water at two or more points in the production process. Though heavily dependent on the nature of the process and the contamination of the wastewater, industry should look at collecting and reusing process wastewater two or more times as the process permits.
3 **On-site treatment:** With the tightening of water supplies and increasing cost of treatment, businesses and industry are finding it may be economical to treat some wastewater on-site for reuse in their facility rather than discharging into the environment. On-site treatment may consist of mechanical or chemical treatment, or natural systems/environmental cleaning as described later in this section.

All wastewater treatment follows the same basic idea. Dirty water that was used in manufacturing, agriculture, municipal uses, food processing, or business activities is contaminated in the production process with chemicals, product waste, particulates, or other contaminants and needs to be cleaned for reuse or discharge.

As defined by Integrated Engineers, Inc. of Oakhurst, California, here are the basic industrial wastewater treatment steps[10]:

1 **Adjust the pH.** Wastewater may become heavily loaded with solids and other contaminants, resulting in pH extremes that can be a problem in later treatment steps. Adjusting the pH allows many wastewater chemicals to work more efficiently later in the process.

2 **Add chemicals.** The addition of flocculating agents begins the process of precipitating solids out of the solution.

3 **Remove contaminants.** Wastewater chemicals break up emulsified oils in water, encapsulate metals and other contaminant, coagulate contaminants, improve pH levels, and separate the solids from the waste stream, allowing easy dewatering. Flocculants are a type of chemical additive that can remove many common industrial contaminants, including:
   a. Phosphates.
   b. Emulsified fats, oils, and grease.
   c. Heavy metals.
   d. Biochemical oxygen demand.
   e. Total suspended solids.
   f. Inks, pigments, and dyes.

4 **Separate liquids and solids.** Removing solids from liquid waste is a key element in the wastewater treatment process. Contaminants contained in the flocculent or sludge are separated from the liquid by removing the solids through the use of a clarifier, a flotation unit, or through screening.

5 **Remove solids.** Sinking solids are removed with a clarifier.
   a. Floating solids are removed with a flotation unit.
   b. Rotary or parabolic screens are used to remove solids that are suspended (neither sinking or floating).

6 **Return water to process.** Once the solids have been removed, the water is returned to a clean condition suitable for reuse in the process.

7 **Biological process.** A biological treatment process can be used after the solid separations phase to remove additional impurities, where necessary, to achieve greater purity for a particular industrial process.

8 **Sludge dewatering.** After completion of the required biological process applications, the wastewater proceeds through the sludge dewatering phase. Sludge dewatering consists of using a compression device to squeeze water out of the waste sludge. It reclaims more recycled water from the process, and leaves the sludge in suitable condition for disposal or reuse. Sludge dryers should be avoided, as they add considerable heating cost to the process.

NATURAL WASTEWATER TREATMENT

Living machines   *Living machines* is a term coined by John Todd, an ecological designer and founder of the Ocean Arks International, a nonprofit environmental organization. A living machine consists of a series of tanks containing plants and other organisms. The combination of elements forms a wetland ecology within a controlled environment, allowing the living machine to more efficiently treat wastewater than natural wetlands. When wastewater is pumped through the tanks, it is naturally treated by the controlled ecosystem contained within the living machine enclosure.

For example, the organisms contain more oxygen than a natural wetland because air is constantly bubbled through the tanks. Some living machines can produce beneficial by-products such as methane gas, and ornamental plants and fish. A study by the EPA, however, of four operating living machines found that they cost about the same as conventional wastewater treatment, and use approximately the same amount of energy. Additionally, living machines located in cold climates must be contained within a heated greenhouse structure, making the economies of the living machine even less desirable. As a visible symbol of a company's commitment to the environment, the living machine has clear value. Unless plant-generated renewable energy or waste heat can be used in its operation, however, the economic case for the living machine remains unproven.

Treatment wetlands   The natural model for the living machine is the wetland. Where plant sites sit on an abundance of land, and local ordinances permit, natural wetlands can be used as an effective wastewater treatment tool. Two conditions must be present for this to work: a sufficient amount of wetland area must be available to accomodate the wastewater volume, and the wastewater must contain a mix of biological agents that can naturally clean the contaminants. The EPA defines a wetland as "an area that is regularly saturated by surface water or groundwater and is characterized by a prevalence of vegetation that is adapted to life in saturated soil conditions" (U.S. EPA, 1994). Wetlands can be divided into two general categories: marshes and swamps. Local jurisdictions may classify some areas as legal wetlands to discourage development even when they do not possess the water-filtering ability of true biological wetlands. Wetlands filter the water by trapping sediment and organic matter. They are often referred to as the "kidneys of the landscape" because of their ability to filter and clean water. Wetlands are desirable habitats for wildlife, and also help control area flooding and lessen drought effects.[11]

Energy Management Systems

Organizations achieving the greatest energy-saving results have:

- A top-down commitment to energy management
- A commitment to continuous improvement
- An approach that integrates energy management across all aspects of the business
- Management systems in place
- A system to regularly assess and track energy performance
- Set measurable performance goals
- An effective reward system for energy performance
- An empowered energy staff

PLANT ENERGY AUDITING

Plant energy audits are comprehensive evaluations of the actual performance of a plant's energy-using systems and equipment compared against the designed performance level or the industry best practice. The difference between observed performance and "best practice" is the potential for energy and cost savings. ENERGY STAR Partners have found that conducting plant audits is vital to a strong energy management program; without audits it is difficult to continuously improve energy efficiency and demonstrate savings.

Energy audits help managers to:

- Identify actions for improving energy performance
- Prioritize projects
- Track progress

Regular plant energy audits are most effective when they are part of a strategic corporate energy management program. Corporate energy programs are ideal for replicating the savings opportunities identified through plant energy audits at other facilities. Through the corporate energy network, information can be shared, and savings multiplied. Read more about the role of plant audits and technical assessments as part of a strategic energy program in the ENERGY STAR Guidelines for Energy Management. www.energystar.gov.

CONDUCTING AUDITS

Energy audits can be self-assessments conducted by company staff, external audits obtained through energy service professionals, or a combination of both.

Regardless of the type of audit, the audit team should include professionals with varied expertise, including process engineers, maintenance experts, systems managers, energy specialists, etc. Support from outside the company is helpful and provides missing expertise and wider industry knowledge that may not be available within the company (such as compressed air systems or refrigeration experts).

Support system review
- Systems that support the process are evaluated, such as compressed air, motors, and steam.
- The review may be conducted by a corporate team supplemented by expertise for major support systems.
- Facilities should budget for slightly higher review costs where outside expertise is required.
- The review should be conducted for a limited time, although it should be longer in duration than a simple walk-through.
- The review should generate specific findings relating to more efficient use of utility systems.
- Ideally, the review results should be usable at multiple plants in a company.
- The review should include sections on lessons learned and best practices shared.

- Managers should recognize that the value of potential savings from a review may generate findings that require no capital investment but do require organizational and operational changes.
- The review may be limited by those systems not covered.
- In manufacturing facilities, the review should include process/yield studies.
- Energy consumption should always be a part of the review.
- The review should include modeling as a critical component to assessing the value of modifications.
- A thorough review can address all major and supporting systems and even impact a company's manufacturing process.

RECOVERY ACT FUNDING

The American Recovery and Reinvestment Act is providing $256 million to support energy-efficiency improvements in major U.S. industrial sectors.[12]

Projects in the following three areas have been selected for award:

1 $156 million for combined heat and power, district energy systems, waste energy recovery systems, and efficient industrial equipment.
2 $50 million for improved energy efficiency for information and communication technology.
3 $50 million for advanced materials in support of advanced clean energy technologies and energy-intensive processes.

ENERGY MONITORING AND TARGETING (M&T)

M&T is an energy-efficiency technique based on the standard management axiom that "you cannot manage what you cannot measure." M&T techniques provide building managers with feedback on operating practices, results of energy management projects, and guidance on the level of energy use that is expected in a certain period.

Energy M&T can be used to compare the relationship of energy use to key performance indicators. More specifically, it can assist building managers in the following goals:

- Identify and explain an increase or decrease in energy use
- Document energy consumption trends (weekly, seasonal, operational. . .)
- Determine future energy use when planning changes in the business
- Identify specific areas of wasted energy
- Observe how the business reacted to changes in the past
- Develop performance targets for energy management programs
- Manage facility energy consumption, rather than accept it as a fixed cost

The ultimate goal of M&T is to reduce energy costs through improved energy efficiency and energy management control. Other benefits generally include increased resource efficiency, improved production budgeting, and reduction in greenhouse gas emissions.

Key M&T principles   M&T techniques rely on main principles, which form a constant feedback cycle, therefore improving control of energy use. See Table 3.7 for a basic list of steps in a typical M&T process.

Monitoring

■ Monitoring is the regular collection of information on energy use, in order to establish a basis for energy management and explain deviations from an established pattern. Its primary goal is to maintain the said pattern, by providing all the necessary data on energy consumption, as well as certain key variables, as identified during preliminary investigation (production, weather, etc.).

Target setting

■ Targeting consists of defining the levels of energy consumption desirable for the management of the facility. Targets are based on previous information obtained during the monitoring phase, as well as detailed knowledge of the business and process requirements. Targeting should be considered desirable objectives for the management and operations teams. They should constitute a challenge, but one that can be reasonably achieved within the available budget and schedule.

Reporting

■ The final principle is the one which enables ongoing control of energy use, achievement of targets, and verification of savings: reports must be issued to the appropriate managers. This in turn allows decision making and actions to be taken in order to achieve the targets, as well as confirmation or denial that the targets have been reached.

Procedures

■ Before the M&T measures are implemented, a few preparatory steps are necessary. First, key energy consumers on the site must be identified. Generally, most of the energy consumption is concentrated in a small number of processes, like heating, or

Tip Box

TABLE 3.7   ENERGY MONITORING AND TARGETING STEPS
1. Monitor collection of data
2. Set reasonable targets
3. Use reliable metering to compile data
4. Define the baseline
5. Monitor variations
6. Identify causes of variations
7. Review results
8. Regularly report results

certain machinery. This normally requires a certain survey of the building and the equipment to estimate their energy consumption level.

■ It is also necessary to assess what other measurements will be required to analyze the consumption appropriately. This data will be used to chart against the energy consumption. These are underlying factors that influence the consumption, often production (for industry processes) or exterior temperature (for heating processes), but may include many other variables.

■ Once all variables to be measured have been established, and the necessary meters installed, it is possible to initiate the M&T procedures.

Measure

■ The first step is to compile the data from the different meters. Low-cost energy feedback displays are available to measure consumption. Models in the market include:
  ■ Powercost monitor
  ■ Cent-a-meter
  ■ The Energy Detective
  ■ Forward Energy Solutions, Inc. [http://www.forwardenergysolutions.com/power.htm]
  ■ Eco-eye [http://www.eco-eye.com/]
  ■ Envi4 [http://www.currentcost.com/]
  ■ Flukso [http://www.flukso.net/]
  ■ PowerWatch [http://www.powerwatch.com/]
  ■ Silk [http://www.silkamr.com/]
  ■ Wattson [http://www.diykyoto.com/uk]
  ■ ECM-1220 [http://www.etherbee.com/BrultechSampleSite]

The frequency at which the data is compiled varies according to the desired reporting interval, but can go from once every 30 seconds to once every 15 minutes. Some measurements can be taken directly from the meters, others must be calculated. These different measurements are often called streams or channels. Driving factors such as production must be collected at intervals to match.

Define the baseline

■ The data compiled should be plotted on a graph to define the general consumption baseline. Consumption rates are plotted on a scatter plot, measured against production or any other variable previously identified, and the best fit line is identified. This graph is the image of the business' average energy performance and documents critical information the manager can use in assessing current performance and potential strategies to save energy.

■ The y-intercept gives the minimal consumption in the absence of the variable (no production, zero degree-day). This is the base load of the system, or the minimal consumption of the system when it is not operating.

■ The slope represents the relationship between the consumption and the previously identified variable. This represents the efficiency of the process.

■ The scatter is the degree of variability of the consumption within operational factors.

- A high y-intercept can mean that there is a fault in the process, meaning it is using too much energy with little improvement in performance. There may, however, be specific aspects of the process that are causing very high base loads. Very scattered points, on the other hand, may reflect other significant factors causing a variation in the energy consumption other than the one plotted initially. These variations can also indicate a lack of control over the process.

Monitor variations

- The next step is to monitor the difference between the expected consumption and the actual measured consumption. One of the tools most commonly used for this is the CUSUM, which is the CUmulative SUM of differences. This consists of first calculating the difference between the expected and actual performances (the best fit line previously identified and the points themselves).
- The CUSUM can then be plotted against time on a new graph, which then yields more information for the energy-efficiency specialist. Variances scattered around zero usually mean that the process is operating normally. Marked variations, increasing or decreasing steadily, usually reflect a modification in the conditions of the process.

Identify causes

- Energy-efficiency specialists, in collaboration with building managers, will decipher the CUSUM graph and identify the causes leading to variations in the consumption. This can be a change in behavior, a modification to the process, different exterior conditions, etc. These changes must be monitored and the causes identified in order to promote and enhance good behaviors, and discourage bad ones.

Set targets

- Once the baseline has been established, and causes for variations in energy consumption have been identified, it is time to set targets for the future. With all this information in hand, the targets are more realistic, as they are based on the building's actual consumption.
- Targeting consists of two main parts: the measure to which the consumption can be reduced, and the time frame during which the compression will be achieved.
- Initially, a good target is the best fit line on the graphs. This line represents the average historical performance. Therefore, keeping all consumption below or equal to the historical average is an achievable and desirable target, yet it remains a challenge because it involves eliminating high energy-consumption peaks.
- As they improve the energy consumption of their facility, some managers may decide to drive their energy consumption down to levels lower than previously achieved. This is a much more challenging goal, obviously, but one that managers can use to focus and energize the company's employees.

Monitor results

- The final step circles back to the original goal: measure consumption. One of the specificities of M&T is that it is an ongoing process, requiring constant assessment in order for a company to consistently improve its performance. Once the targets are set and the desired measures are implemented, repeating the procedure from the start ensures that managers can assess the effectiveness of each measure, giving them the information to help decide on what further action to take.

Endnotes

1. U.S. Department of Energy. "Introduction to Commercial Building Control Strategies and Techniques for Demand Response." http://drrc.lbl.gov/pubs/59975 .pdf. Accessed August 12, 2010.
2. Sezgen, O. and J. Koomey. 1998. Interactions between Lighting and Space Conditioning Energy Use in U.S. Commercial Buildings. Lawrence Berkeley National Laboratory. LBNL 39795. Berkeley, CA.
3. U.S. Department of Energy, Federal Energy Management Program. "Tips to Conserve Natural Gas." http://www.msbo.org/services/energy/ReduceNaturalGasUse.pdf. Accessed March 7, 2010.
4. Seco/Warwick Corporation. "Metal Minutes: Aluminum Heat Processing News." http:// www.secowarwick.com/metalminutes/maintenancetips/algasatmosconsumption.htm. Accessed September 15, 2010.
5. Alliance for Water. "Water Smart Guidebook (Process Water)." http://www.alliance-forwaterefficiency.org/uploadedFiles/Resource_Center/Library/non_residential/ EBMUD/EBMUD_WaterSmart_Guide_Process_Water.pdf. Accessed February 18, 2010.
6. U.S. Energy Information Administration. http://www.eia.doe.gov/cneaf/solar .renewables/page/trends/rentrends.html. Accessed March 7, 2010.
7. Law360.com. "Utilities And Renewable Energy: To Buy Or To Build?" http:// www.troutmansanders.com/files/upload/RenewableEnergy_ToBuyortoBuild.pdf. Accessed September 15, 2010.
8. U.S. Department of Environmental Protection Green Power Partnership. "Renewable Energy Certificates." http://www.epa.gov/greenpower/documents/gpp_basics-recs .pdf. Accessed March 14, 2010.
9. U.S. Environmental Protection Agency. "On-site Renewable Energy Generation." http://www.epa.gov/statelocalclimate/documents/pdf/7.2_on-site_generation.pdf. Accessed August 12, 2010.
10. Integrated Engineers, Inc. http://www.wecleanwater.com/html/services/waste-water-treatment-process.htm. Accessed September 15, 2010.
11. Stovall, Heather. "Natural Alternatives to Conventional Wastewater Treatment." http://lda.ucdavis.edu/people/2007/HStovall.pdf. Accessed August 14, 2010.
12. U.S. Department of Energy Industrial Technologies Program. http://www1.eere .energy.gov/industry/index.html. Accessed March 7, 2010.

4

OPERATIONS

The rule of thumb among building managers is that the total cost of a facility can be broken down as follows: 10 percent construction, 90 percent operations and maintenance. Whatever the actual breakdown for a given building, it is clear that controlling facility operational costs is the best route available to a manager to minimize costs and maximize profits.

Energy and other utility costs are the largest single component of operational costs. Reducing energy consumption is obviously a priority for any facility manager, but numerous other opportunities exist across the full range of building operations to save money and operate in a more sustainable manner. Part of sustainability also includes maintaining a safe and healthy workplace free of hazardous air, chemicals, and materials.

Indoor Vehicles

Forklifts are the workhorses of the industrial environment. They provide horizontal and vertical transportation for a wide variety of products. Common forklifts can transport up to 4000 pounds (1814 kilograms) of products on shipping pallets, can lift up 20 feet (6.1 meters) in height, are easily maneuverable, and are low-maintenance.

Internal combustion engines (Classes 4–8) were traditionally powered by gas, diesel, compressed natural gas, or liquid propane. These fuels have fallen out of favor because of increasing concerns with carbon monoxide emissions and indoor air pollution, and their long-term effects on warehouse and industrial personnel.

Electric vehicles (Classes 1, 2, 3, and 6) have become popular for indoor applications because their performance rivals that of internal combustion engines without the environmental concerns.

For the most part, these trucks are used indoors because they produce no harmful emissions.

See Table 4.1 for a table of forklift classes and codes.

TABLE 4.1   FORKLIFT CLASSES AND LIFT CODES

CLASS 1: ELECTRIC-MOTOR RIDER TRUCKS

- Lift Code 1: Counterbalanced Rider Type (Stand-up)
- Code 4: Three-Wheel Electric Truck (Sit-down)
- Code 5: Counterbalanced Rider Type (Cushion Tires, Sit-down)
- Code 6: Counterbalanced Rider Type (Pneumatic or Cushion Tire, Sit-down)

CLASS 2: ELECTRIC-MOTOR NARROW-AISLE TRUCKS

- Lift Code 1: High-Lift Straddle
- Code 2: Order Picker
- Code 3: Reach-Type Outrigger
- Code 4: Side Loaders, Turret Trucks, Swing-Mast, and Convertible Turret/Stock Pickers
- Code 6: Low-Lift Pallet and Platform (Rider)

CLASS 3: ELECTRIC-MOTOR HAND TRUCKS

- Lift Code 1: Low-Lift Platform
- Code 2: Low-Lift Walkie Pallet
- Code 3: Tractors (Draw-Bar Pull, Under 999 lb)
- Code 4: Low-Lift Walkie (Center Control)
- Code 5: Reach-Type Outrigger
- Code 6: High-Lift Straddle
- Code 7: High-Lift Counterbalanced
- Code 8: Low-Lift Walker/Rider Pallet

HYDROGEN FUEL CELLS

Hydrogen fuel cells offer the potential to provide power for plant vehicles in a lower-cost, safer, and more environmentally friendly manner. Nuvera Fuel Cells recently developed a power product called PowerEdge™. Based on the longtime PowerFlow™ fuel cell system, this product replaces standard lead acid batteries in forklifts.

PowerEdge is a hybrid system consisting of a fuel cell power system; sealed, maintenance-free batteries; and a compressed hydrogen storage system. Unlike conventional batteries, hydrogen fuel cells stay on the forklift and can be quickly refueled, allowing operators to get back to work in only a few minutes. PowerEdge, for instance, is a smart system, equipped with remote monitoring capabilities to allow managers of large facilities to better manage their forklift fleet.

HYBRID FORKLIFT SYSTEMS

Another approach to increasing forklift efficiency is to combine existing technology with a means of quickly charging the batteries on board the vehicle.

Exide Technologies and Ballard Power Systems are teaming up to develop hybrid power systems for forklifts. Unlike most fuel cell systems for cars that have used either nickel metal hydride or lithium ion batteries, Exide is using lead acid battery packs as the primary power source. They will be combined with a Ballard's Mark 1020 fuel cell stack, which has been formerly used mostly for stationary applications, such as telecom backup systems. This hybrid system, though still based on common battery technology, is designed to be lower cost. It provides a cost-effective solution to keeping batteries charged on the forklifts so that banks of replacement batteries do not have to take up plant floor space.

Indoor Chemicals

CHEMICAL CONTAMINANT SOURCES

There are a variety of chemical contaminants found in a variety of sources. Volatile organic compounds (VOCs) are common chemical contaminants found in office and home environments and are a source of odors. VOCs are organic (containing carbon) chemicals that can easily evaporate into the air. Many products found in the office environment may have the potential to release VOCs. Examples include[1]:

- Adhesives
- Caulks, sealants, and coatings
- Paints, varnishes, and/or stains
- Wall coverings
- Vinyl flooring
- Cleaning agents
- Air fresheners and other scented products
- Fuels and combustion products
- Carpeting
- Fabric materials and furnishings
- Personal products of employees like perfume and shampoos

BUILDING PRODUCT EMISSIONS

See Table 4.2 for a list of potential employer problems resulting from indoor chemicals.

Indoor hazards
- Bioaerosols from water damage, microbial VOCs (VOCs from fungi)

Tip Box

TABLE 4.2   EMPLOYER INDOOR CHEMICAL RISKS
Employee health problems
Employee absenteeism
Reduced employee productivity
Worker's compensation claims
U.S. Environmental Protection Agency (EPA) or Occupational Safety and Health Administration (OSHA) complaints.
Employee litigation
New employee recruitment difficulties
Diminished employer reputation

- Emissions from building carpet, furnishings, and other building components (VOCs including formaldehyde from glues, fabric treatments, stains, and varnishes)
- Emissions from office equipment (VOCs, ozone)
- Emissions from stored supplies (solvents, toners, ammonia, chlorine)
- Emissions from special use areas within the building, such as laboratories, print shops, art rooms, smoking lounges, beauty salons, food preparation areas, and others (various chemicals and related odors)
- Emissions from indoor construction activities (VOCs from use of paint, caulk, adhesives, and other products)
- Accidental events such as spills inside the building
- Elevator motors and other building mechanical systems (solvents and other chemicals)
- Emissions from pesticide use inside the building
- Plumbing problems (sewer odors, improper bathroom ventilation)
- Emissions from housekeeping/cleaning activities (ammonia, chlorine, and other cleaning agents such as detergent, dust residual from carpet shampoo, and disinfectants)
- Use of deodorizers and fragrances
- Emissions from stored trash inside the building
- Fire damage inside the building (soot, polychlorinated biphenyls from electrical equipment, odors)

However, recent research also shows associations between respiratory/allergic health effects and indoor concentrations of chemicals, as well as common indoor materials and finishes. Indoor factors associated with health effects include specific organic compounds, formaldehyde plasticizers, aromatic compounds, aliphatic compounds; and indoor finishes or materials such as particleboard, flexible flooring, plastics, paint, carpet, and renovation activities.

Formaldehyde risk factors   Higher formaldehyde concentrations or presence of particleboard increases the risk factors for respiratory problems, particularly for employees with these conditions:

- Diagnosed asthma
- Adverse changes in lung function
- Diagnosed chronic bronchitis
- Lung inflammation
- Wheeze
- Atopy or allergy

Plastic finishes risk factors   Higher dust concentrations of phthalates (BBzP, DEHP) or presence of plastic surfaces may increase the following health effects:

- Diagnosed asthma
- Bronchial obstruction
- Wheeze, cough, phlegm
- Allergy, exzema, rhinitis

Paint and dust risk factors   Recent painting or renovation may increase these health effects:

- Wheeze or obstructive bronchitis
- Pulmonary infection
- Allergy

Facility manager actions   To prevent health problems associated with chemical exposures in the workplace, owners and managers should:[2]

- Respond to building-related health concerns of workers.
- Establish clear procedures for recording and responding to indoor environmental quality (IEQ) complaints.
- Ensure an adequate and timely response to all complaints.
- Record all complaints by date.
- Collect detailed information about each complaint, taking care to document the specific health effects.
- Ensure confidentiality to the person lodging the complaint.
- Craft a plan for correction of the problem and responding to the complainant.
- Identify appropriate resources to use in correcting the problem.
- Correct the situation that caused the health concern. Where a specific correction (or a specific problem) cannot be identified, seek to remediate the effects on the individual through other means.
- Provide frequent responses to building occupants regarding the complaint and the actions taken in response to it.
- Follow up with the complainant to ensure that remedial action has been performed and was effective.

- Repair any areas of water incursion or leakage. Verify no mold or damp material remains.
- Schedule building renovation work, such as interior painting and installation of new carpets or wall coverings, after work hours or when the building is unoccupied.
- Open windows or increase building ventilation to dilute chemical odors on a temporary basis. Remove odor-causing source for a long-term solution.
- Choose low-emitting, low-VOC office supplies and cleaners.
- Choose low-VOC or zero-VOC emitting products when choosing new or replacement carpets, flooring, wall coverings, office furniture, and paints.
- Ask facility suppliers for product information on chemical emissions and potential health hazards.
- Use pesticides reluctantly, and apply them only when the building is vacant. Follow the integrated pest management system to prevent risks of exposure.
- Use U.S. Environmental Protection Agency (EPA) information on the use of nonchemical methods of pest control, such as baits or traps, whenever possible. http://www.epa.gov/pesticides.
- Provide proper ventilation and maintain heating, ventilating, and air conditioning (HVAC) systems.
- Properly store cleaning and maintenance chemicals with containers closed and tightly sealed.
- Do not store chemical products in equipment rooms where they could contaminate the HVAC system.
- Do not block air distribution diffusers with office equipment or furniture.
- Do not use concentrated products at full strength. Dilute products to their recommended strength before using.
- Ensure that the manufacturer's instructions for the use of all cleaning products are followed.
- Do not allow workers to mix different types of cleaning products. Combinations of different cleaners may give off poisonous fumes.
- Ensure that all cleaning product label precautions are followed. Keep all cleaning products in their original containers.

Water Management and Recycling

COMMERCIAL CONSERVATION TIPS

- **Educate.** Educate employees, contractors, and other users in each facility about the importance of conserving water.
- **Survey.** Perform a leak survey on the facility. Observe the meter after hours (or on weekends) to see if there are any leaks. Larger or more complex facilities may require submetering to help identify leaks.
- **Plant smart.** Plant low-water-use plants or native grass instead of non-native grasses or water-intensive plants.
- **Low-water.** Replace older toilets and urinals with high-efficiency toilets and urinals. Many utilities offer rebates for both residential and commercial toilet replacements. Waterless urinals are now acceptable in many municipalities.

- **Low-flow.** Replace old aerators and showerheads with new low-flow types.
- **Maintain.** Regularly inspect, maintain, and repair facility boiler systems. Install a condensate return line on all boilers.
- **Identify.** Identify all single-pass flows in the facility. These flows often are associated with equipment cooling for pumps, compressors, ice machines, air conditioners, and other equipment.
- **Inspect.** Regularly inspect, maintain, and repair the building's cooling tower system, which consumes a significant amount of a building's water. Some utilities have a cooling tower audit program.
- **Review.** Review historical water usage for the facility. Analyzing several years of consumption data will often identify undiscovered leaks or other problems.
- **Audit.** Perform a water audit on the facility to help identify where water goes, and which appliances or pieces of equipment are the most intensive water users.
- **Eliminate.** Eliminate single-pass (once-through) cooling water in air compressors and other equipment by using chillers, cooling towers, or air-cooled equipment. Minimize blowdown on cooling towers and boilers by using a conductivity controller.

XERISCAPE PRACTICES

Xeriscape practices use low-water-consuming plants to create a landscape that is sustainable and less expensive to maintain. Denver Water coined the word in 1981 as a tool to help make low-water-use landscaping an easily recognized concept. Xeriscape is a combination of the word *landscape* and the Greek word *xeros*, which means dry. When properly designed, Xeriscape projects can provide dense, attractive landscaping that is substantially less costly to water and maintain. The Xeriscape concept relies on seven principles:

Sustainable planning and designing, limiting turf areas, selecting and grouping plants appropriately, improving the soil through additives, using mulch effectively, irrigating efficiently, and maintaining the existing landscape as much as possible. See Table 4.3 for LEED New Construction credits available for using Xeriscape and other sustainable landscaping practices.

SOIL AMENDMENTS

Amending site soil with compost retains moisture in the soil, making water available to plants for longer periods of time. Compost provides small amounts of important plant nutrients, including nitrogen, phosphorus, and potassium, all of which improve root growth. The additives also open up clay soils for better drainage and close sandy soils to prevent water from leaching away from the surface (and plant roots) too quickly.

Amendment recommendations:

- Remove rock and debris larger than 1 inch (2.5 centimeters) in diameter to avoid interfering with planting and maintenance.
- Add compost soil amendments to all permeable areas of the property, including tree lawns and right-of-way land the owner is responsible for maintaining.

TABLE 4.3   LEED EXISTING BUILDING CERTIFICATION LANDSCAPING REQUIREMENTS

LEED New Construction, Water Efficiency Credit 1.1
Water Efficient Landscaping: 1 point
Reduce landscaping water usage by 50%.

LEED New Construction, Water Efficiency Credit 1.2
Water Efficient Landscaping: 1 point
No potable water use or no irrigation.

LEED Existing Buildings, Water Efficiency Credit 1.1
Water Efficient Landscaping: 1 point
50% reduction in potable water use for irrigation over conventional
  means of irrigation.

- Depending on original soil conditions, amend up to 1000 square feet (93 square meters) of soil with 4 cubic yards (3.1 cubic meters) of approved compost.
- For greatest effect, incorporate compost into the soil to a depth of 4 to 6 inches (10.2 to 15.2 centimeters).

SPECIFIC INDUSTRY STRATEGIES[3]

Water policy
- Include water efficiency in the company's environmental policy statement.
- Create an employee manager with water-efficiency responsibilities.
- Use a submeter and monitor to identify excessive water consumption or leaks.
- Inspect all equipment and supply lines for leaks regularly.
- Establish water-efficiency goals and communicate those goals to all employees.

Hotel guest rooms
- Install low-flow or high-efficiency toilets.
- Install sink aerators to reduce the flow to a maximum of 1.5 gallons (5.7 liters) per minute.
- Install aerators on mop sinks to reduce usage to no more than 2.5 gallons (9.5 liters) per minute.
- Install showerheads with a flow rate of 2.5 gallons (9.5 liters) per minute or less.
- Install urinals that use less than 1 gallon (3.8 liters) per flush.
- Install water closets that use 1.27 gallons (4.81 liters) per flush.
- Begin a linen reuse program for hotel customers.
- Equip all ice machines with air-cooled condensers.

See Table 4.4 for an *ROI Quik-Calc* for replacing water closets (toilets) with lower water-consuming models.

ROI Quik-Calc

TABLE 4.4 WATER-CONSERVING TOILET REPLACEMENT
Current gallons per flush (GPF) average = 1.60 GPF (6.0 liters per flush)
Average GPF savings after replacement with water-efficient unit: Assume: 1.27 GPF (4.81 liters per flush)
Water savings per flush after replacement: 0.33 GPF (1.25 liters per flush)
Estimate average annual fixture usage = 39,000 flushes
50 users per toilet / 3 times per day / 260 days per year
Annual water savings per fixture: 39,000 * 0.33 = 12,870 gallons (48,718 liters)
Estimated water/sewer rate cost per 1,000 gallons (3,785 liters) of water: $7.00
Annual savings per toilet: $91
Estimated cost of toilet replacement: $180
Investment cost recovery: 180 / 91 = 2 years approx.
Return on investment (ROI) over 5 years: (450 / 180) = 250%

Restaurants

- Notify customers that the restaurant serves water only upon request.
- Post notices in the restaurant food-preparation and dishwashing areas to encourage employees to conserve water. Stress water conservation at staff meetings.
- Use a food-service checklist that includes information about conserving water in restaurants.
- Defrost food in the refrigerator instead of under hot water.
- For ice-making, use energy-efficient flake or nugget machines instead of cube ice machines.
- Utilize less than 15 gallons (56.8 liters) of water in combination ovens (typically used to keep food from drying out while baking).
- Use kitchen steamers (boilerless type) that consume 3 gallons (11.4 liters) per hour or less.
- Use boiler-type steam kettles to collect the condensate from steam, with an insulated line to help the water retain heat. Reuse the water for boiling.
- Use pasta cookers with a simmer mode and automatic overflow controls.
- Install an in-line restrictor on the dipper well supply line for the purpose of reducing flow to less than 0.3 gallons (1.1 liters) per minute.
- Install 1.6 gallons (6.1 liters) per minute prerinse spray valves to rinse dishes.
- Install a water-saver kit on waste grinders or use strainer baskets in place of garbage disposers.
- Use dishwashers and C-line conveyors with electronic sensors and door switches to stop water flow when the dishwashers are not in use.
- Replace worn and missing dishwasher water jets.

- Use high-efficiency dishwasher equipment and install steam doors to reduce evaporation.
- Install dishwashers that retain water from the last wash to use in the next load (do not use fill-and-dump machines).

Janitorial
- Use hose flow at less than 5 gallons (18.9 liters) per minute, along with the use of self-closing nozzles.
- Use rubber squeegees to collect food residue from the floor prior to using a hose.
- Sweep floors rather than hosing them down with water.

Restrooms and plumbing
- Install low-water-usage toilets.
- Install urinals that use 1 gallon (3.8 liters) of water per flush or less. Consider water-less urinals.
- Install aerators on hand-washing sinks to reduce each sink's flow to 1.5 gallons (5.7 liters) of water per minute or less.
- Install aerators on the kitchen, food service, and mop sinks to limit the flow rate to less than 2.5 gallons (9.5 liters) per minute.
- Install showerheads with a water flow rate of 2 gallons (7.6 liters) per minute or less.

Heating and cooling
- Eliminate single-pass (once-through) cooling water used in air compressors through the use of air-cooled equipment, chillers, or cooling towers.
- Minimize blowdown on cooling towers and boilers by using a conductivity controller.

Irrigation and landscape
- For irrigation purposes, do not exceed 18 gallons of water per square foot (40.7 liters per square meter) of irrigated areas.
- Install drought-resistant or native landscaping plants.
- Minimize areas of turf on the site. Maximize areas of native plants that can exist on normal area rainfall.

Commercial car wash
- Reduce sprayer nozzle sizes and water pressure to a maximum flow rate of 3 gallons (11.4 liters) per minute. Check for leaks frequently and repair quickly.
- Discontinue bay/lot wash downs.
- Install a wash-water reclamation system.
- Do not use more than one soap pass.
- Replace brass or plastic nozzles with stainless steel nozzles for longer life. Plastic and brass nozzles wear away faster, increasing the nozzle size and allowing increased water flow over time.
- Check nozzle alignment for automatic systems regularly.

■ Install a temperature-controlled weep management system and set it to 32°F (0°C).
■ Eliminate the spot-free rinse cycle.
■ Increase the speed of wash cycles.

Equipment Efficiency

See Table 4.5 for common industry efficiency strategies.[4]

MOTOR SYSTEMS EFFICIENCY

There are two general methods for saving energy in motors: ensuring that the motor is highly efficient, and matching the motor power efficiently to the demand. Implementing an effective motor strategy involves carefully considering the necessary loads and designing an efficient motor strategy to satisfy those loads. The results are likely to save energy, demand charges, and maintenance costs, as well as improve productivity.

The most basic and reliable strategy to save the cost of motor operation is to replace low-efficiency motors with high-efficiency motors. Over its lifetime, the cost of a motor can be outstripped by the cost of the energy it uses by a factor of 100 or more. For this reason, an improvement in motor efficiency of only a few percent is usually a cost-effective investment for a facility as it will recover the cost of the upgrade within a short period. Use the following strategies in reducing facility motor loads:

1 Match motors with loads. Failure to match motors with loads is a leading cause of needless electrical energy consumption. In many industrial, agricultural, and

Tip Box

TABLE 4.5   COMMON INDUSTRY EFFICIENCY MEASURES
HVAC: Increase efficiency
HVAC: Reduce operating hours
HVAC: Reduce load
Lighting: Supplement with daylighting
Lighting: Reduce fixture load
Lighting: Install energy-efficient fixtures
Water: Reclaim and reuse process water
Water: Install low-water-use fixtures
Water: Use drought-resistant landscaping
Motors: Match motors to load
Motors: Begin motor maintenance plan

commercial applications, motors are oversized for most of the time they are in use. Installing a variable speed device that allows the motor to run as efficiently as possible for the instantaneous needs of the task at that moment can be a particularly cost-effective retrofit. This is a good solution for motors that move fluids, such as pumps, fans, and air compressors. Cutting back on motor power to the point where flows are just adequate for the load saves considerable energy. In addition, installing multiple pumps, fans, or compressors and staging their operations to match loads is another practical energy- and cost-saving strategy.

2 **Launch a motor maintenance program.** A motor maintenance program should include routine inspections of all motors, with an emphasis placed on those critical to production. The inspection procedure should include checking the drive train, measuring energy use, and identifying any overheating mechanical and electrical components.

AIR COMPRESSOR EFFICIENCY

Most plants use compressed air for at least some manufacturing or production functions. For many facilities, compressor energy costs form a substantial portion of their entire electric bill. Many compressor systems are poorly configured, contain leaking fixtures, and include motor and compressor systems that are mismatched to their loads. Suggestions for curbing energy waste in air compressor systems include the following:

- Regularly maintain the system to minimize air leakage.
- Optimize the mechanical design of the system, using a closed loop system if practical. Larger pipes result in less mechanical (frictional) losses, allow operation at reduced pressure, and cut down on peak demand.
- Ensure that systems can operate efficiently at partial loads.
- Use properly sized, energy-efficient compressors driven by energy-efficient motors, with storage tanks that are matched to loads.
- Design the system to minimize air leakage. Perform regular maintenance.
- Use electronic controls on individual compressors if applicable.
- Meter energy, flow, and other parameters to assess system performance and minimize system air pressure.

Reducing air leakage in a compressed air system is very important. For a compressed air system in constant operation in a facility whose electric energy cost is 5 cents/kWh, a $1/8$-inch-diameter (3.175-millimeter) leak costs about $2000 per year, while a $1/4$-inch-diameter (6.35-millimeter) leak can lose approximately $8000 per year.

Lighting

Lighting is responsible for approximately 9 percent of total electricity use in the industrial sector. There is large potential for cost-effective lighting energy savings in many industrial facilities. Economy measures frequently found to be practical include the following:

- Paint ceilings and sidewalls with a white semigloss paint. This will enhance the lighting quality at most work stations by raising brightness levels and softening shadows and glare whether light is from electric fixtures or from the sun.
- Consider replacing conventional high-intensity discharge lighting in medium and high bays with fixtures that use more efficient T-5 fluorescent lamps that may be dimmed stepwise when daylighting is available.
- Replace T-12 fluorescent fixtures with T-8 or T-5 fixtures with electronic ballasts.
- Install systems that redirect direct sunlight from high windows onto light-colored ceilings, reducing glare and converting sunlight into a supplementary lighting source.
- Install and adjust automatic dimming controls to take advantage of daylighting. The "Cool Daylighting" approach keeps most outside light out of the field of view, thereby controlling for glare, producing better distribution, and lowering cooling costs. See www.daylighting.org/what_is_cool_daylighting.htm for additional information.
- Install energy-saving light-emitting diode (LED) exit signs.
- Upgrade parking lot lighting to save energy and reduce the environmental impacts associated with lighting the sky instead of the parking surface.

INDUSTRY-SPECIFIC STRATEGIES

Grocery store
HVAC recommendations

- Install a demand-controlled ventilation (DCV) system. When only a few people are in a store, energy can be saved by decreasing the amount of ventilation supplied by the HVAC system. A DCV system senses the level of carbon dioxide in the return air stream and uses it as an indicator of occupancy. DCV can save energy during peak cooling periods when many shoppers are at work and occupancy is low.
- Install variable air volume air handling systems with variable speed drives.
- Install high-efficiency motors.
- Choose high-efficiency packaged air conditioning units.
- Downsize to high-efficiency chillers in conjunction with lighting and refrigerator case retrofits to units that produce less heat and leak less cool air into the store interior. Sizing the HVAC equipment to take into account the cool air leaking from cases and cabinets can usually justify downsizing the chiller, offsetting the higher first cost of high-efficiency equipment.
- Recommission the building's HVAC, compressed air, and refrigeration systems at least once a year.
- Use condensing boilers with large turn-down ratios whose efficiencies improve with lower loads.
- Switch HVAC systems to direct digital controls.

- Upgrade the facility's energy management system and optimize settings to better reflect building usage, respond to changing outside temperatures, and control peak electric loads.
- Consider using evaporative cooling in dryer climates with high heat and lower humidity.

Refrigeration

- It is usually worthwhile to upgrade refrigeration systems in grocery stores to include efficient, state-of-the-art technologies. Such upgrades should include installing dew point controls for anticondensate heaters on refrigerated cases so defrosting is matched to actual need.
- Incorporate efficient cooling system components, such as high-efficiency compressors; water-cooled condensers; floating-head pressure controls; and multiple, unequally sized compressors feeding the same manifold.
- Install floor insulation in coolers. The floors of some walk-in refrigerators in many grocery stores are simply concrete slabs that are neither insulated from the earth underneath nor around their edges. Retrofitting coolers with floor insulation improves their efficiency.
- Install rolling night covers over open freezer units. See Table 4.6 for an *ROI Quik-Calc* for savings associated with this improvement.
- Use efficient lighting in refrigerators and save twice: this retrofit lowers the electricity use for both lighting and cooling.

Hotels and motels
Hot water

- Inspect frequently for water leaks and repair them quickly. Ignoring such simple maintenance measures is costly since leaks tend to get worse with time and become more expensive to fix.
- Install low-flow showerheads. Models whose spray patterns may be adjusted by users are best as they communicate to guests that management cares about both comfort and energy/water conservation. Payback periods are very rapid on low-flow showerheads in hotels.
- Lower hot water system temperature to between 120 and 130°F (49 and 54.4°C).

ROI Quik-Calc

TABLE 4.6 GROCERY STORE OPEN FREEZER NIGHT COVERS
Assume freezer length of 32 feet (10 meters)
Electric consumption of freezer without night cover = 1700 kWh per day
Electric consumption of freezer with night cover = 1650 kWh per day
Calculate savings per day (50 kWh * number of freezers * electricity cost per kWh)
Calculate annual savings: Savings per day * 365
Savings / investment = Return on investment (ROI)

- Install a wastewater heat recovery system to preheat hot water.
- Insulate hot water supply and recirculating lines.
- Use a hot water recirculating system to minimize water usage.
- Install gas-fired, high-efficiency hot water heating equipment. Small atmospherically vented water heating systems with energy factors of 0.62 to 0.70 are more cost-effective than standard equipment with lower efficiencies. Direct vent, sealed-combustion, condensing boilers possess energy factors up to 0.86.
- Install multiple boilers in new installations. Multiple boilers provide redundancy and can be staged in a way that more efficiently meets the hot water load than a single large boiler.

HVAC

- Install occupancy controls for lighting and HVAC in guest rooms.
- Assess the benefit of heat-pump water heaters for indoor swimming pools to simultaneously heat water and provide dehumidification.
- Institute on-demand ventilation controlled by air quality sensors located in specific public spaces throughout the facility, including hotel lobbies, dining rooms, and parking garages.
- Choose high-efficiency packaged air conditioning units.
- Downsize to a new high-efficiency chiller in conjunction with lighting retrofits.
- Use condensing boilers with large turn-down ratios whose efficiencies improve with turn-down.
- Consider using evaporative cooling systems in warmer/dryer climates where it may be effective.
- Switch system to direct digital controls.
- Install high-efficiency motors.
- Verify proper operation of economizer functions and controls.
- Assess the benefit of using cool air from the cooling tower with water-cooled chillers.
- Upgrade the building's energy management system, and optimize its settings to reflect actual usage, outside temperature fluctuations, and peak electric loads.
- Insulate all steam piping, hot air ducts, and hot water piping to reduce losses. See Fig. 4.1 for a photograph of insulated rooftop process steam piping.

Lighting

- Install compact fluorescent bulbs in place of incandescents in guest rooms, halls, and elevators.
- Install energy-saving LED exit signs
- Install energy-efficient lighting in all other spaces.
- Install and calibrate automatic lighting controls in conjunction with skylights and clerestories in open areas in order to dim lights in response to daylight.
- Upgrade parking lot lighting to save energy and reduce environmental impacts due to light spillage.
- Upgrade garage parking lighting to LED or high-efficiency lighting.

Figure 4.1   **Photograph of insulated rooftop
process steam piping.**

HVAC/Power

■ Investigate installing a combined heat and power (CHP) system. Hotels and motels
are good applications for such systems because in addition to supplying all of the
site's electrical loads, a CHP system can supply all of the site's heat and hot water
for laundries, showers, food service, and space heating. When properly sized and
designed, this system can save substantial amounts of money, avoid the large thermal losses associated with conventional power generation at utility plants, avoid the
transmission and distribution losses associated with delivering the power over power
lines, and avoid separate fuel usage for heating.

■ As part of the CHP system, incorporate an absorption chiller into the design. An
absorption chiller can run off the waste heat from a CHP system, drastically cutting
down on peak-time electricity used for cooling loads.

Warehouses
Lighting

■ Install photo sensors and occupancy controls to control electric lighting, and make
sure they are carefully calibrated for safe lighting levels.

- Replace T-12 fluorescent fixtures with more efficient T-8 or T-5 fixtures with electronic ballasts.
- Include skylights with photocell controls into the facility to provide daylighting and reduce lighting electrical loads from conventional lighting. Place skylights above warehouse aisles to achieve best daylighting distribution.
- Consider replacing metal halide and low-pressure sodium fixtures with T-5 fixtures.
- Provide task lighting controlled by occupancy sensors in narrow aisles to more efficiently illuminate tall storage racks. Mount fluorescent fixtures on storage racks to facilitate access to the storage aisles and avoid the need to install extra fixtures at the ceiling level.

HVAC

- Control heating, ventilating, and cooling systems as a function of occupancy and the needs of goods stored using automatic controls as necessary.
- Install gas-fired infrared heaters instead of forced-air convection heating systems.
- Use variable speed drives, high-efficiency motors, and cast aluminum fan blades for ventilation fans, and use demand control to adjust ventilation rates as needed.

Retail stores
Lighting

- Replace T-12 fluorescent fixtures with T-8 or T-5 fixtures with electronic ballasts. Bulbs with good color rendering properties for displaying merchandise are widely available.
- Use higher-efficiency spot and flood lighting for illuminating merchandise. Incandescent and halogen lighting are both inefficient and produce considerable heat. In many cases, these types of bulbs can be replaced by compact fluorescent fixtures that provide better illumination, are much more efficient, and last from 4 to 10 times longer than incandescent and halogen fixtures.
- Install LED exit signs.
- For new construction or renovations, consider installing skylights and photocell controls to gain daylighting savings. In many retail designs, standard 4-foot by 8-foot (1.2-meter by 2.4-meter) square skylights with bubble-style caps are used. Usually, a skylight-to-floor area ratio of 1:25 balances daylight with space conditioning requirements. Energy savings and enhanced sales can result from skylight installation.
- Install and adjust automatic dimming controls to take advantage of daylighting.
- Upgrade parking lot lighting to save energy and reduce environmental impacts.

HVAC

- Install a demand-controlled ventilation system. When only a few people are in a store, energy can be saved by decreasing the amount of ventilation supplied by the HVAC system. A DCV system senses the level of carbon dioxide in the return air stream and uses it as an indicator of occupancy. DCV can save energy during peak cooling periods when many shoppers are at work and occupancy is low.

- Choose high-efficiency packaged air conditioning units.
- Install variable air volume air handling systems with variable speed drives.
- Downsize to a new high-efficiency chiller in conjunction with lighting and other retrofits.
- Use condensing boilers with large turn-down ratios whose efficiencies improve with turn-down.
- Switch system to direct digital controls.
- Install high-efficiency motors properly sized to the load.
- Upgrade the building's energy management system, which will enable savings from settings that better reflect usage, changing outdoor temperatures, and interior load requirements.
- Recommission building systems frequently.

Operations and Maintenance

Most commercial rooftop HVAC systems have a life span of approximately 15 years. Maintaining the unit on a planned preventative maintenance schedule, or even replacing a unit earlier then absolutely necessary, can substantially reduce energy usage and help avoid lost revenue by eliminating years of the cost of decreased efficiency and rising maintenance and repair expenses.

ENERGY STAR-rated HVAC systems offer significantly higher Seasonal Energy Efficiency Ratio (SEER) efficiency ratings than equivalent-sized equipment of even five years ago. According to the EPA, the owner of a commercial building can recover $2 to $3 of incremental asset value for every $1 invested in energy performance improvements. Many utility companies also offer rebates for installing higher-efficiency HVAC equipment.

Commissioning

Building commissioning is the process of systematically checking the performance of major building systems to ensure they are installed and operating in conformance with the manufacturer's specifications, and that the performance of the systems is meeting the requirements of the design professionals and building owners.

Lennox Corporation offers their Prodigy™ unit controllers for their Strategos™ and Energence™ rooftop units. This controller includes a guided setup and intuitive interface to help technicians and installers ensure that system components are installed in accordance with the design specifications. This may simplify the job of a hired commissioning agent, but will not eliminate the need for such an agent in satisfying the commissioning requirements for LEED certification.

Cleaning

When seeking green chemicals for commercial, industrial, or business applications, keep the following features in mind:

- Materials should be readily biodegradable
- VOC free
- Nontoxic
- Plant-based
- Noncorrosive and nontoxic
- Chlorine-free and phosphate-free

Industrial cleaning solutions fall into two broad categories: chemical detergents and industrial organic chemicals. Chemical detergents are synthesized using artificial chemicals. These are powerful, but contain toxic substances. Industrial organic chemicals, on the other hand, are equally powerful, but do not contain toxic substances. Organic chemicals are sourced from plants and vegetables. These products do not contain artificial or harmful substances. As such, the cleaning efficiency of green chemicals does not depend on a single hazardous chemical; instead these formulas use advanced green technologies to achieve the cleaning power.[5]

The most powerful green chemicals make use of a technology that involves extremely small, nano-sized particles that filter into stains and dirt residues and break the bond between the surface and dirt. Resulting particles are encapsulated and emulsified in water for quick and easy removal. See Table 4.7 for LEED Existing Building credits available for sustainable building cleaning practices and products.

ADVANTAGES OF GREEN CHEMICALS

The main advantage green chemicals offer is the reduction of health problems in workers. The workers in most industrial facilities and factories work in demanding conditions. Artificial cleaning solutions contain many toxic substances that can cause disorders and diseases, such as cancer, reproductive deficiencies, respiratory illnesses, and skin disorders. The use of green chemicals poses reduced risk of health problems for users, and consequently reduced liability and worker lost time for employers.

Additionally, the use of organic cleaners is a smart move for industrial facilities looking to reduce their carbon footprint, as these formulas are nontoxic, readily biodegradable, and safe for the environment.

Carpet extractors and cleaning solutions are also used for cleaning the interior of vehicles, especially carpets, mats, and seat upholstery. Specialized products have been developed for various aspects of cleaning automotive interiors. Green versions of these specialized detailing products are available, including for such purposes as mobile car washes, windshield glass cleaners, carpet shampoos, and upholstery cleaning solutions.

TABLE 4.7 LEED EXISTING BUILDING CERTIFICATION GREEN CLEANING REQUIREMENTS

Indoor Environmental Quality Credit 10.1

Entryway Systems: 1 point

Install entryway grills, grates, mats to reduce dirt or dust particles entering the building. Implement cleaning strategies for entryways and exterior walkways.

Indoor Environmental Quality Credit 10.2

Isolation of Janitorial Closets: 1 point

Install or renovate a completely enclosed closet with: outside exhaust, no air recirculation, negative pressure, hot and cold water, and appropriate chemical drains.

Indoor Environmental Quality Credit 10.3

Low Environmental Impact Cleaning Policy: 1 point

Implement a low-impact environmental cleaning policy that addresses sustainable cleaning: systems, policies, personnel training, waste disposal, and other attributes.

Indoor Environmental Quality Credit 10.4 – 10.5

Low Environmental Impact Pest Management Policy: 1-2 points

Implement a low-environmental-impact indoor pest management policy. The policy must require use of cleaning products that meet the requirements identified in LEED MR credit 4.1–4.3.

Indoor Environmental Quality Credit 10.6

Green Cleaning: Low Environmental Impact Cleaning Equipment Policy: 1 point

Implement a cleaning policy using janitorial equipment that minimizes contaminants and environmental impact. Upgrade existing janitorial equipment as required.

Cleaning detergents were previously the norm for industrial cleaning because earlier versions of organic cleaning products lacked sufficient cleaning power. Because cleaning jobs such as heavy-duty industrial degreasing and component washing require strong cleaning agents, it was believed that organic substances were not strong enough to do the job.

However, today's top green chemicals are as powerful as, or even more powerful than, their artificial counterparts. At present, a variety of green chemicals suitable for various types of heavy-duty cleaning jobs are available on the market at competitive prices to standard industrial products. The improvement in IEQ resulting from the use of organic cleaning products should lead to less employee illness and employer liability from disability or workplace allergy claims.

EPA DESIGN FOR THE ENVIRONMENT STANDARDS

EPA's Design for the Environment (DfE) program works in partnership with environmental groups, industry, and academia to reduce risk to people and protect the environment by finding ways to prevent pollution. For over 15 years, DfE has used partnership projects to evaluate traditional and alternative chemicals in a range of industries for their effect on human health and the environment.

When the DfE logo appears on a product, it means that the DfE review team has screened each ingredient in the product for potential human health and environmental effects and that—based on hazard and risk information, the latest models and predictive tools, and expert scientific judgment—the product contains only those ingredients that pose the least concern among chemicals in their class. For example, if a DfE-recognized product contains a surfactant, then that surfactant will not be toxic to human health and it will biodegrade readily to nonpolluting degradation products; many surfactants in conventional products biodegrade slowly or biodegrade to more toxic and persistent chemicals, which threaten aquatic life.

DfE allows manufacturers to put the DfE label on household and commercial products, such as cleaners and detergents, that meet stringent criteria for human and environmental health. Using these products can protect human health and the environment.

The program contains two tools that can be of benefit to building managers:

- The Alternatives Assessments Program helps industries choose safer chemicals by providing an in-depth comparison of potential human health and environmental impacts of the chemicals they currently use or plan to use. This tool is suitable for situations in which safer chemicals have yet to be identified as viable substitutes.
- The Best Practices Program shows workers how to protect themselves and their communities' health by using chemicals safely and minimizing exposures.

The DfE program is unique because of two defining characteristics: its assessment methodology and its technical review team. The DfE technical review team has many years of experience and is highly skilled at assessing chemical hazards, applying predictive tools, and identifying safer substitutes for chemicals of concern. The review team applies the DfE assessment methodology by carefully reviewing every product ingredient (the review includes all chemicals, including those in proprietary raw material blends, which manufacturers share with DfE in confidentiality).[6]

Shifts and Scheduling

FLEXTIME

The most common time-management options for employers are variable work hours (also called flextime) and compressed workweek schedules. Flextime programs, in which employees are given the option of adjusting their arrival and departure times,

TABLE 4.8   EMPLOYEE FLEX TIME STRATEGIES

- Ten-hour workdays over four days
- Work from home one day per week
- Mini-shifts to provide coverage for other time zones
- Staggered work hours to reduce commuting time
- Work from home computer connections

are proven ways to reduce congestion at peak travel times. Compressed work schedules (such as a four-day, 10-hours-a-day workweek) can eliminate one day a week of commuting for many employees.

Advantages of flextime and compressed work schedules

- Help implement changes companywide or by department
- Increase coverage for companies that communicate across time zones for customers requiring extended hours
- Allow employees to travel to and from work with less stress during off-peak hours
- Provide employees with greater flexibility in planning medical and personal appointments
- Enable employees to choose to work during their most productive hours (flextime)

Programs that give employees the option of adjusting their arrival and departure times from work help to avoid peak travel times and reduce traffic during the worst period of the day. Another effective way to help cut commuting time is a compressed work schedule (such as a four-day, 40-hour workweek), which can eliminate one day of commuting per week.

Other schedule-saving strategies are:

- Implement 10-hour-days, with one day off.
- Allow workers to work at home one day per week.
- Stagger work hours for employees to avoid traveling during peak traffic times.

See Table 4.8 for a tip box for company employee flextime strategies.

Carpooling and Transportation

The average U.S. household uses 1143 gallons (4327 liters) of gas per year. It is expected that the number of cars and trucks on already crowded highways will double in the next 30 years. The average American spends 434 hours (18 days) in his or her car each year. The average worker will drive their car 20,800 miles (33,474 kilometers) a year and emit over 23,600 pounds (10,705 kilograms) of carbon dioxide (CO_2). Every car annually emits its own weight in CO_2. Cutting 25 miles (40 kilometers) a week from an employee's driving distance can save 1500 pounds (680 kilograms) of CO_2. The United States could save 33 million gallons (124,920 kiloliters) of gas each day if the average commuting vehicle carried one additional person.[7]

The average passenger car emits (or consumes) the following each year:

■ 80 pounds (36.3 kilograms) of hydrocarbons
■ 606 pounds (275 kilograms) of carbon monoxide
■ 41 pounds (18.6 kilograms) of nitrogen oxides
■ 10,000 pounds (4,536 kilograms) of carbon dioxide
■ Uses 550 gallons (2,082 liters) of gas

The average sport utility vehicle (SUV) emits (or consumes) the following in a year:

■ 114 pounds (52 kilograms) of hydrocarbons
■ 894 pounds (406 kilograms) of carbon monoxide
■ 59 pounds (27 kilograms) of nitrogen oxides
■ 17,000 pounds (7,711 kilograms) of carbon dioxide
■ Uses 915 gallons (3,464 liters) of gas

CARPOOLING BENEFITS

Carpooling is defined as two or more people commuting to work together on a regular basis in a privately owned vehicle. It is the simplest and most common form of ridesharing. At a workplace employees may choose to carpool without any assistance or involvement from the employer; however, carpool incentive programs are a way for employers to encourage employees to carpool. Carpool incentive programs may incorporate a variety of means to encourage employees to carpool. See Table 4.9 for a short list of employer and employee carpooling/rideshare benefits.

Possible incentives include reduced cost or free parking, preferred parking, or reward programs (such as prize drawings). Other rideshare benefits are as follows:

■ Employers can help employees form carpools through rideshare matching, which helps potential carpoolers locate others nearby with similar schedules. Regional rideshare organizations in most areas allow interested employees to register directly for no cost. Employers can direct their employees to these free services.
■ Employee benefits from carpooling include cost sharing, less wear and tear on vehicles, time savings in regions with high-occupancy vehicle (HOV) lanes, and the ability to talk, eat, sleep, or read while commuting.
■ The principle employer benefit of ridesharing is the need for fewer employee parking spaces. Other secondary advantages include reduced employee stress and improved productivity.

TABLE 4.9   BUSINESS CARPOOL/RIDESHARE BENEFITS

Employer Benefits

    Less employee absenteeism

    Low-cost employee benefit

    Reduced parking load

    Works with flextime and scheduling strategies

    Possible regional rideshare resources

Employee Benefits

    Reduced commuting expenses

    Reduced commuting frustration

    Potential tax benefits from savings or employer assistance

    Assistance with arranging carpool/rideshare partners

    Potential employer benefits (preferred parking/flextime)

■ Programs to encourage carpooling, such as rideshare matching services, preferred parking for carpools, and reduced parking costs for carpools.

When carpool incentive programs make sense
■ Regions with HOV lanes.
■ Employers with limited parking.
■ Employers with large numbers of employees.
■ Employers in urban settings.
■ Tax provisions that allow carpool parking costs to be taken as a tax-free fringe benefit, offering potential financial savings for both employers and employees.

Tax benefits
Commuting benefits may be provided tax-free to employees each month up to amounts allowed by the U.S. Internal Revenue Service (IRS) (currently over $200 per month). Tax benefits accrue to businesses and employees whether the employer pays for the benefits or the employee pays for it through a pretax salary deduction. If parking costs are less than the allowed amount, parking benefits can only equal the actual cost of parking. However, any employee who drives to work is eligible for these benefits, not just carpoolers.

IMPLEMENTATION ISSUES AND COSTS

With rideshare matching, one of the most important needs in setting up a carpool program is matching potential carpool partners who are compatible with each other in terms of location and working hours. Outside some obvious situations (a married couple who both work for the same employer), potential carpoolers may not know anybody with whom to carpool.[8]

Preferred parking   One workplace incentive for carpoolers is preferred parking. Spaces can be designated for either individual carpools or rideshares in general.

Reduced-cost parking   Where employers charge for parking, reducing the parking cost for carpools encourages carpooling.

Employee schedules   A lack of matching schedules among employees can be a major barrier to carpooling. Senior management can encourage carpooling by requesting that supervisors take carpoolers' needs into account. This includes predictable schedules and infrequent requests to work overtime.

Fraud potential   Employers with carpool programs report that employees sometimes commit carpool fraud. There are several sources of potential fraud, including registration of carpool partners who do not actually carpool, self-reporting of carpool trips when the driver actually drove alone, and creation of false carpool passes or tags.

Other incentives   In areas where parking is free and plentiful, employers can encourage carpooling through prize drawings or other rewards.

Services that support implementation   Many regions have rideshare organizations whose primary function is to maintain databases of potential carpoolers. Because of both the difficulty of keeping this type of database updated, and the fact that many employers do not have sufficiently high numbers of employees to make many carpool matches, a regional rideshare service is an excellent way to find carpool partners for employees.

Parking cash out   Parking cash out is an arrangement in which an employer offers employees cash in lieu of a parking space.

Emergency ride home programs   A barrier preventing some employees from carpooling is the fear that they will not be able to get home quickly in the event of an emergency, such as picking up a sick child from school or working unscheduled overtime. Emergency ride home programs provide commuters, who regularly carpool, vanpool, bike, walk, or take transit to work, with a reliable ride home when an unexpected emergency arises.

Park-and-ride lots   For potential carpool partners who do not live in immediate proximity to each other, a park-and-ride lot may be a good meeting place. Park-and-ride lots may be set up by municipalities on public or private property in some locations. More often, individual carpool partners will need to make arrangements with local shopping center or development owners to park in outlying areas of their lots.

Other factors affecting transportation programs

- Regional ridesharing databases can help in identifying carpool partners.
- Ride matching services offered by agencies or the company improve participation rates.
- A guaranteed ride home service can relieve the anxiety of transportation worries resulting from car mechanical troubles or expenses.
- Less wear and mileage on employee vehicles is a powerful selling feature.
- Less stressful commute/less driving is a benefit for both employer and employee.
- Reducing traffic congestion and fuel consumption is a benefit to society.
- Regional benefits of ridesharing, including helping to keep the air clean.
- Unexpected benefits also occur, like making new friends and keeping more consistent work hours.

Where carpooling or rideshare programs are not appropriate or workable for an employer, the employer may consider employee inducements for the following:

- Public transportation
- Vanpooling
- Bicycling

EMPLOYEE CARPOOLING TIPS

Offer employees these guidelines to get the most out of their carpool arrangement:

- Rideshare participants should contact everyone on the potential match list.
- Call all commuters who live and work or attend school near them and have similar work or school hours, even if they have not selected the category the employee is looking for (e.g., drive only, share driving, ride only).
- Negotiate with prospective rideshare partners to see how much flexibility they have.

Cover the basics   In making carpool or rideshare arrangements, employees should agree on answers to some key questions:

- How many people should be in the rideshare group?
- How often will each member rideshare?
- Who in the group owns a vehicle? If all passengers have a car, will each rotate driving?
- Do all rideshare drivers have full insurance coverage?
- Where will the rideshare group meet? Carpoolers may pick each other up at home, meet at one residence, or meet at a mutually convenient location such as a park-and-ride lot.
- When will the rideshare group meet in the morning? Rideshare partners' work or school schedules may or may not be flexible enough to accommodate each other.

- What will be the ridesharing costs? If commuters rotate the driving equally, money may not need to be exchanged. If only one person drives in the rideshare, passengers should contribute to cover the costs of gas, tolls, and parking.
- Do all rideshare partners know one another? Carpoolers who do not know each other sometimes feel more comfortable meeting prospective carpool partners before they drive together for the first time. Members should plan to talk on the phone or meet at a public place to discuss carpool specifics and decide whether or not they would feel comfortable sharing a ride. If a member still feels uncomfortable after meeting, he/she can simply choose not to pursue the rideshare arrangement. No one should be obligated to carpool. If one party does not feel comfortable enough to agree to a long-term rideshare arrangement, perhaps that person could agree to a trial rideshare period to test the waters.
- What are the group's rules? Each rideshare member should have a chance to express his or her needs and concerns. Carpoolers should agree on certain ground rules at the outset:
  - Agree on food, coffee, smoking, and perfume/cologne usage
  - Agree on radio or music choices
  - Agree on how long drivers will wait for delays
  - Determine who is notified if someone is sick
  - Agree on driving safety requirements
- Should the group give carpooling a trial run? Many commuters start carpooling on a trial basis, for a month or two. A person can always add more days and more carpoolers in the future once a routine has been established. It is not usually necessary to get the details perfect right away.

Rideshare resources   The best resources for companies seeking assistance in setting up ridesharing programs can often be found with local governmental or planning agencies. They will usually possess the most detailed and area-specific information on how best to organize a corporate program, and where to find local assistance, rideshare databases, and how to link with public transportation programs. National online resources include:

Rideshare: www.rideshare.com

Rideshare Directory: www.rideshare-directory.com/

E-Rideshare: www.erideshare.com/

Continuous Improvement

Continuous Improvement Process (CIP, or CI) is an ongoing effort within a business to improve products, services, or processes. These efforts can seek incremental improvements over time or breakthrough improvement all at once. Delivery (customer valued) processes are constantly evaluated and improved in the light of their efficiency, effectiveness, and flexibility.

Some see continuous improvement as an overarching process that can be an umbrella for most management systems, such as business process management, quality management, or project management practices. Deming saw it as part of the "system," whereby feedback from the process and customer were evaluated against organizational goals. The fact that CIP can be called a management process does not mean that it needs to be executed by management, merely that it makes decisions about the implementation of the delivery process and the design of the delivery process itself.

Some successful CIP programs use the approach known as Kaizen [the translation of kai ("change") and zen ("good")]. This method became famous in the book of Masaaki Imai, *Kaizen: The Key to Japan's Competitive Success*. This term signifies a daily activity, the purpose of which is to streamline normal work practices and help people find and eliminate waste in their activities. People at all levels of an organization can participate in Kaizen. This practice involves a small group cooperating to improve their own work environment and productivity. While Kaizen usually delivers small improvements, the culture of continual, aligned small improvements and standardization yields large results in the form of compound productivity improvement.

Components of most CI and Kaizen programs are as follows:

- The core principle of CIP is the (self) reflection of processes. (Feedback)
- The purpose of CIP is the identification, reduction, and elimination of suboptimal processes. (Efficiency)
- The emphasis of CIP is on incremental, continuous steps rather than giant leaps. (Evolution)
- Improvements are based on many small changes rather than the radical changes that might arise from research and development (R&D).
- As the ideas come from the workers themselves, they are less likely to be radically different, and therefore easier to implement.
- Small improvements are less likely to require major capital investment than major process changes.
- The ideas come from the talents of the existing workforce, as opposed to using R&D, consultants, or equipment—any of which could be very expensive.
- All employees should continually be seeking ways to improve their own performance.
- It helps encourage workers to take ownership for their work, and can help reinforce team work, thereby improving worker motivation.

The elements above are the more tactical elements of CIP. The more strategic elements include deciding how to increase the value of the delivery process output to the customer (Effectiveness) and how much flexibility is valuable in the process to meet changing needs.

A systems approach has been developed to address preventive actions often referred to as pollution prevention or cleaner production. This approach is presented in the EPA publication, "An Organizational Guide to Pollution Prevention."[9]

Construction Activities

Economic health is a key aspect of sustainable development. According to the EPA, it is "the ability to achieve continuing economic prosperity while protecting the natural systems of the planet and providing a high quality of life for its people." Or, as defined in 1987 at the United Nations World Commission on Environment and Development, sustainable development "meets the needs of the present without compromising the ability of future generations to meet their own needs."

Sustainable commercial construction uses resources efficiently to contribute to the health of the environment and its inhabitants. At the same time, these buildings should provide economic return to the firms that own them through superior energy performance and reduced waste. These high-performance buildings may use gray water or harvested rainwater to irrigate, be located within walking distance of public transportation, and have designated areas for recycling and composting waste, among other measures. They all take into account the five key aspects of green building: site sustainability, water efficiency, energy efficiency, enhanced IEQ, and sustainable materials. The goal is to reduce consumption and pollution from cradle to grave—from the building's construction to its ultimate deconstruction.

The most recognizable green building rating system in the United States, the LEED system of the U.S. Green Building Council (USGBC), has rating systems for various building projects, including LEED for New Construction, LEED for Existing Buildings, and LEED for Commercial Interiors.

LEED for Existing Buildings is designed to rate older buildings that are retrofitted for resource efficiency.

LEED is just one certification system. Sustainable commercial construction buildings may also receive ratings from other third-party verification systems, like the Green Building Initiative's (GBI) Green Globes system, or earn a qualification from the EPA's ENERGY STAR program.

Steps to building a sustainable commercial space   The economic benefits of green building in the commercial space are often debated, but analysis has shown that while high-performance buildings may cost more to construct than conventional buildings—because of the cost of some sustainable building materials and green building consulting fees—they more than make up for those expenses in reduced energy and water bills and operations and maintenance costs.

Federal tax credits also exist for green building projects, and according to the Green Outlook Report published by McGraw-Hill Construction, the number of states with green building policies, standards, legislation, and programs increased from 13 to 31 between 2005 and 2008.

SUSTAINABLE CONSTRUCTION

The success of any green building project—whether new construction, retrofit, or reuse—relies on the goals a company and its design professionals set, the team they assemble, the tools they use, and the systems they install to evaluate performance. When embarking on such a project, companies should be sure to do the following:

1 **Set goals.** Establish goals for the sustainability of the building, and make sure they are measurable. Learn from the successes of other high-performance buildings.
2 **Create an integrated design and construction team.** From the outset, obtain the expertise of a collaborative team of architects, engineers, designers, builders, and sustainable construction specialists, and solicit feedback from the building's occupants, owners, and maintenance staff. With this holistic approach, managers are more likely to see early cost savings.
3 **Take advantage of industry tools and resources.** Modeling software, guides for how to save on resource consumption, including guidelines for green building certification from LEED or other third-party evaluators, are all valuable aids in the process of building green.
4 **Measure and evaluate building system performance.** Once the building is constructed and the systems are in place, regularly monitor and periodically assess the results to ensure the systems are operating efficiently. Make changes as required to keep them performing at peak levels.

Endnotes

1. Centers for Disease Control and Prevention. "Indoor Environmental Quality." http://www.cdc.gov/niosh/topics/indoorenv/ChemicalsOdors.html. Accessed February 20, 2010.
2. Indoor Chemical Emissions and Respiratory Health of Children. Mark Mendell, Ph.D., Indoor Environment Department, Lawrence Berkeley National Laboratory. "Respiratory Health of Children." http://www.iaqsymposium.com/pdfs/presentations/Building_A_Firm_Foundation_Mendell.pdf. Accessed January 10, 2010.
3. Denver Water. "Conservation." http://www.denverwater.org/Conservation. Accessed March 7, 2010.
4. Energy Efficiency Guide for Colorado Businesses. "Recommendations by Sector." http://www.coloradoefficiencyguide.com/recommendations/default.htm. Accessed August 12, 2010.
5. Green Chemicals Guide. http://www.greenchemicalsguide.com/. Accessed March 7, 2010.
6. U.S. Environmental Protection Agency. Design for Environment Standards. http://www.epa.gov/dfe/. Accessed August 7, 2010.
7. Rideshare Company. "The Rideshare Company will Save the Planet." http://www.rideshare.com/employer_environmental_benefits.html. Accessed September 15, 2010.

8. U.S. Environmental Protection Agency. "Carpool Incentive Programs: Implementing Commuter Benefits as One of the Nation's Best Workplaces for Commuters." http://www.gpoaccess.gov/harvesting/carpool.pdf. Accessed September 15, 2010.

9. Pojasek, Robert. "Once is not Enough: Continual Improvement is Essential to Sustainability." http://www.greenbiz.com/blog/2008/12/29/once-not-enough-continual-improvement-essential-sustainability. Accessed January 10, 2010.

5

BUILDING FEATURES

Automatic Entry Doors

Automatic entry doors are sliding, swinging, revolving, or folding doors that open and close automatically when a sensor indicates a person is approaching the door. They are most often found in retail and supermarket occupancies, where large numbers of people need to access the building and standard swing doors represent a significant liability risk and inconvenience for customers. Automatic openers are increasingly being installed to meet local barrier-free accessibility requirements and to comply with the Americans with Disabilities Act (ADA). Depending on the type of installation and door speed, automatic doors can provide significant energy savings by reducing air infiltration and heat loss. Doors and building spaces in the proximity of doors, such as vestibules and other spaces and rooms, must comply with local accessibility requirements, most of which are based on the ICC Standard A117.1 Accessible and Usable Buildings and Facilities and on Americans with Disabilities Act Accessibility Guidelines (ADAAG).

AUTOMATIC ENTRY DOOR STRATEGIES

Sliding doors   One of the most commonly used types of automatic doors is sliding doors. Sliding doors offer superior energy performance, saving up to 50 percent of the energy of automatic swinging doors and double the savings of manually operated swinging doors. Automatic doors are operated by either a floor mat sensor or a detector mounted directly above the door. Opening and closing times are widely adjustable depending on the type and customer characteristics of the business. In industrial environments, automatic doors offer the advantage of hands-free operation for areas within a facility where doors are required for thermal, environmental, or other reasons. The hands-free operation of sliding doors means slightly greater productivity in saving the time to open and close doors, particularly when transporting material. See Table 5.1 for an *ROI Quik-Calc* for converting a standard storefront door to an automatic sliding door.

One of the largest detriments of retrofitting existing facilities to automatic sliding doors has always been the width required to accommodate the installation. Conventional

ROI Quik-Calc

TABLE 5.1   AUTOMATIC ENTRY DOORS

Assumption: Conversion of a single 6-foot-wide conventional pair of swing doors to a 12-foot-wide automatic sliding door unit.

Cost: $6000 (typical) – $10,000 (structural modifications required)
New installation = $4000 – $6000

Return on Investment: 8–10 years

Note: ROI does not include customer convenience or throughput factors.

sliding doors require an installation width that is double that of the required door opening. This requirement can pose a severe hardship in tight building areas. Newer designs, however, employ a "telescoping door" strategy that allows sliding doors to operate in widths of slightly more than 50 percent larger than the door opening itself.

Revolving doors   Revolving doors are available in a range of sizes, including diameters up to 20 feet (6.1 meters), and consisting of 2–4 turning vanes. Because of their efficiency in saving energy, revolving doors in high-usage situations provide a return on investment (ROI) of 3–5 years. Revolving doors must be paired with other door types for egress purposes.

Swing doors   Swing doors are less energy efficient than sliding or revolving doors, mainly because of the additional time required for them to operate, allowing more air infiltration losses. Swing doors, which are available from various manufacturers (most notably Nabco, Inc. Acugard), can operate on 12–24 VDC and 24 VAC low-energy power sources. Building managers can adjust hold-open times to reduce air infiltration losses. Gradually reducing the hold-open times of swing doors over a period of time will slowly accustom building users to moving through the opening faster, but may also result in more door maintenance expenses.

Folding doors   A folding door is an accordion-style automatic door that offers the energy-saving benefits of an automatic door system without the width requirements of sliding door systems or the depth requirements of swing systems. Opening and closing times for folding doors are greater than those of either swing open or sliding doors, and they tend to require more maintenance and are more subject to user damage than other styles. Folding doors occur in two styles: pairs of doors usually occur in a bifold arrangement, or as a single unit.

Vestibules and Airlocks

A vestibule is a small entrance to a building, separated by two doors and usually having no other function than to provide a separation between a building's interior spaces and the outdoors. An airlock is essentially the same, except the term is used mostly

ROI Quik-Calc

TABLE 5.2   VESTIBULES AND AIRLOCKS
Assumption: Installation of a prefabricated exterior vestibule with lighting
Cost: $8000 – $12,000 per unit
Return on Investment: 7–9 years

in conjunction with industrial facilities where it is specifically intended to separate a particular building environment (clean room, paint booth, or freezer, for example) from the normally conditioned building environment. Both vestibules and airlocks offer the benefits of providing a barrier to air infiltration into heavily-used pedestrian buildings, either through a large volume of people or repetitive use by employees in moving between spaces frequently. Vestibules are now required by many building and energy codes for access to public lobbies and other spaces. The code usually stipulates a minimum distance between doors (typically no less than 7 feet), and sometimes a minimum vestibule floor area as well. See Table 5.2 for an *ROI Quik-Calc* for adding an exterior vestibule to a building.

Vestibules in industrial applications are useful even when they are not required by the building code. Any access, service, or plant door that is frequently used by employees can result in significant infiltration loss in the winter (and excess heat gain in the summer). Prefabricated vestibule units are available for secondary doors such as these, and can be purchased and installed in a variety of configurations. Plant managers and designers must carefully consider whether barrier-free provisions of the code apply before installing a vestibule in these types of locations.

Where vestibules are not appropriate for either space or cost purposes, exterior shelters can provide some of the same benefits by screening the sides of the doors from most direct winter wind and creating a partial "dead air" zone around the door opening. Although their benefits are not as great as those provided by a full vestibule, their cost is significantly less.

Overhead Doors

Because of their size and perimeter area, overhead doors allow a significant amount of air infiltration into facilities. Though of lesser concern in warehouses or other buildings that can tolerate wider temperature fluctuations, air infiltration is a serious issue in commercial facilities where even a small number of overhead doors can make working conditions in the loading area uncomfortable. In supermarket service areas, where produce and perishable items may require some amount of conditioning, leaky overhead doors can be a significant source of energy loss and additional utility costs. Overhead doors are used mainly in exterior applications, but are often installed in interior applications to separate cooler or freezer areas from unconditioned portions of a facility or

Figure 5.1   Energy-efficient overhead door.

other areas that require only normal conditioning. See Fig. 5.1 for a photograph of an energy-efficient overhead door with head and jamb seals.

Industrial buildings have traditionally used a handful of basic strategies to reduce energy losses at overhead doors:

INSULATED DOORS

Standard practice (often incorporated in energy codes) is to install doors with a minimum of 3 inches (7.6 centimeters) of insulation (providing an insulation value of approximately R-12). Older industrial facilities may have overhead doors with minimal or no insulation, and suffer excessive conductive losses as a result.

DOCK SEALS

See loading docks below for an explanation of the benefits of dock seals.

EDGE BAFFLES OR SEALS

Flexible neoprene edge baffles or brush seals are effective means of reducing air infiltration around door perimeters. Brush seals are as effective as neoprene, but typically offer more durability and longer service.

DOOR BOTTOM SILLS

Door sills, also referred to as bottom thresholds, are critical to controlling air infiltration at overhead doors. These usually take the form of compressible bulbs made of butyl or other durable vinyl material. They serve the twin purpose of reducing air and water infiltration under the door. Door sills are more subject to damage and deterioration than head or jamb components, and should be inspected and replaced when torn or deteriorated.

HIGH-SPEED DOORS

High-speed doors operate at speeds up to 60 inches (152 centimeters) per second. Although high-speed door systems normally possess effective edge seals to reduce air infiltration, because they consist of monofilament flexible panels, they have virtually no insulation. They are most effective in situations where a door may be opened frequently and it is desirable to minimize the amount of time spent opening and closing it, such as interior clean room or freezer separations.

ENVIRONMENTAL CONSIDERATIONS

Where overhead doors separate interior areas of a building with special environmental concerns, such as clean room environments, paint booths, or areas where chemical treatments are applied, perimeter protection in the form of sill and side jamb protection is especially critical to limit dust or chemical vapor penetration through the opening. Door systems with special coatings and seals are usually specified to match the needs of the particular application. Durability and ease of replacement of the door seals are major considerations in special applications such as these.

Loading Docks

Sustainable strategies for loading dock improvements include:

LED DOCK LIGHTS

Low-energy light-emitting diode (LED) dock lights use approximately 25 percent of the energy of a typical halogen dock light. When coupled with a motion detector to automatically cut the light off when not in use, dock light energy usage can be substantially reduced. In large distribution or truck transfer applications, dock light energy reduction can result in surprisingly large overall energy savings.

DOCK SEALS

Dock seals are vertical compression pads that mount on each side of an overhead dock door. They are extremely effective at sealing the side openings when a trailer is backed up to the dock for loading or off-loading. Custom angled seals are available for depressed docks where the truck is backed up to the dock via a sloping ramp. The ROI of dock seals depends on the amount of time the dock is utilized and the temperature difference between indoors and outdoors. Typical paybacks on dock seal installations, however, run from 3 to 5 years. They are a very cost-effective way to achieve energy efficiency in a facility with frequent dock access.

DOOR WEATHER STRIPPING

See the overhead door section in this chapter for a discussion of doors.

DOCK LEVELERS

Dock levelers are often ignored as a source of heat loss and infiltration in industrial buildings. While dock seals and shelters seal the jambs and head opening at overhead doors, gaps at the dock leveler are, for the most part, ignored. This "fourth" opening surrounding overhead doors can be a major source of heat loss even when a truck is not using the dock. To make dock levelers less inefficient, building managers can install an under-leveler pit seal to provide a barrier when the leveler is not being used. This pit seal consists of a compressible vinyl curtain and other elements that block gaps when the leveler is in the lowered position. When the leveler is raised (in the above-dock position), a separate header curtain drops to maintain the seal. The ROI for this installation is 1–2 years.

Lighting

The illumination of interior spaces is one of the most energy-intensive aspects of operating any commercial or industrial facility.

BASIC LIGHTING COST-SAVING STRATEGIES

Lighting efficiency   One method of reducing electric costs is to use energy-efficient fixtures or bulbs (see Chap. 8 for a listing of resources). Offer incentive programs to business for upgrading office or warehouse lighting to achieve greater efficiency.

Fewer fixtures or bulbs   A time-honored technique in offices is to remove fluorescent tubes from overhead fixtures, reducing a four-tube fixture to a two-tube fixture, for instance. This strategy does save money, but not as much as replacing the low-efficiency fixture with a more efficient model. A four-tube fixture, for instance, has a ballast sized for the full wattage of the fixture. If two of the four bulbs are removed, the lighting output certainly drops in half, but the energy consumption does not since the ballast provided with the fixture continues to draw up to 75 percent of the energy consumed under when all four tubes are in place. A secondary factor is that the optics of the fixture are designed around a specified number of bulbs. Simply removing bulbs, particularly in fluorescent ceiling fixtures, results in variations in the ambient lighting of the room, with more dark/light fluctuation. Deleting bulbs from an existing fixture is an energy-saver, to be sure, but it comes with penalties in other areas and should be considered as a short-term expedient, with the long-term solution to be the installation of more efficient lighting sized for the space.

Task lighting   In industrial facilities, particularly those where fabrication or manufacturing processes are occurring, it is frequently inefficient to raise the ambient lighting of the entire production area up to the required level. Common wisdom held that high ambient levels were necessary to accommodate the frequent changes in industrial equipment locations and work areas, allowing for maximum flexibility. Though a reasonable ambient level of lighting throughout the space is necessary for efficient and safe working conditions, task- or area-specific lighting may be more desirable to accommodate specific work areas.

LED lights   LEDs represent a major advance in lighting technology. Used in fixtures as diverse as street lights, traffic signals, and interior lighting, LEDs offer superior energy performance and long life. Expense and lumen output have limited widespread adoption of LED technology in building lighting, but improvement in both of these areas will make LED lighting an important tool for reducing lighting costs in the future. See Fig. 5.2 for a photograph of LED bulbs.

Daylighting   Daylighting strategies (the use of indoor natural light) can be used to reduce the need for artificial light (see section on daylighting) during both sunlit and cloudy periods of the day.

ADVANCED LIGHTING COST-SAVING STRATEGIES

Flexible configuration   Advanced lighting control systems, such as Encelium Technologies' Energy Control System (ECS) Greenbus technology, allow light fixtures to be connected via communication cables and special connectors. Through graphic software, building users can control their lighting environment "on the fly" by easily changing light controls from one switch to another, or changing a bank of lights to be controlled by occupancy and photo sensors. This system also allows managers to tune in response to daylighting, personal needs, demand response (load shedding), time scheduling, or occupancy.

LIGHTING CONTROL SYSTEMS

Lighting control systems enable all the components of a lighting system—photosensors, switches and other controls, ballasts, and occupancy sensors—to communicate. These systems vary widely in capability, but most allow a manager to control lighting across his facility down to a circuit of fixtures, and even to individual fixtures. Savings are achieved by allowing sensors to automatically shut off lighting when natural light is available or when no employees are occupying a space. Some systems can adjust the lighting level in a space to preset levels, dimming lighting in warehouse areas, for instance, during periods when they are rarely accessed.

Figure 5.2   Light emitting diode (LED) bulbs provide more light with less electricity.

WAREHOUSE LIGHTING

Warehouse lighting systems can reduce lighting costs, save energy, and enhance productivity throughout an entire warehousing operation. These systems are specifically engineered to maximize kilowatts to light narrow aisles, medium-width aisles, and wide open staging and assembly spaces. This type of system reduces energy consumption and maintenance costs. Other advantages are as follows:

- It is a more desirable system for lighting rack aisles because the narrow aisles and very tall racks do not impede correctly designed lighting systems.
- This system reduces lighting overhead costs.
- High-efficiency lighting provides excellent ROI, reducing maintenance costs and lowering power bills while actually improving light levels.

- The reflector design focuses light only where it is needed, maximizing the foot candles per watt consumed.
- Fixture line employs high-lumen fluorescent T5H0 lamps, which maintain up to 95 percent of lumen output over their 20,000-hour-rated life compared to the 33 percent loss of light at just 8000 hours for high-intensity discharge metal halide lamps.
- Compared to other fluorescent fixture manufacturers, focused warehouse aisle systems can save 50 percent in energy costs while maintaining the same light levels.

Daylighting

Daylighting is the process of using natural light to replace much of the need for artificial lighting in a facility. Daylighting is provided through the controlled admission of natural light into an industrial or commercial interior space through windows or skylights. The National Institute of Building Sciences estimates that the use of daylighting can reduce building energy costs by as much as one-third. See Fig. 5.3 for a photograph of the roof of a warehouse facility using skylights for daylighting purposes.

Figure 5.3 Warehouse facility skylights provide energy-saving daylighting benefits.

BENEFITS OF DAYLIGHTING

Increased employee productivity   The use of natural daylight in a work environment has been shown to increase employee satisfaction in their work environment and connection to the outdoors. It also leads to modest increases in productivity.

Reduced emissions   Because daylighting reduces the need for electric consumption for lighting and cooling, the use of this strategy reduces greenhouse gases and fossil fuel depletion.

Reduced operating costs   Typical electric lighting is responsible for 35 to 50 percent of the total electrical energy usage in commercial buildings. Because electric light fixtures also generate heat, their use adds to the building's mechanical equipment loads for cooling. For this reason, the energy savings realized from using daylighting strategies in lieu of electric lighting can directly reduce building cooling energy usage by 10 to 20 percent. For many institutional and commercial buildings, total energy costs can be reduced by as much as 33 percent through the optimal use of daylighting strategies.

Improved life-cycle cost   Because daylighting can save in fixture, wiring, and utility costs, its use can dramatically reduce the installed cost of new facilities, saving as much as 20 percent on the lighting and electric subcontract. Savings over the lifetime of the facility are even more impressive, as annual savings from utilizing daylighting strategies can save up to $0.30 per square foot ($3.22 per square meter) of operating area in lighting-intensive areas.

Implementing daylighting strategies in industrial facilities is accomplished through the use of skylights and any number of roof or wall glazing configurations. Increasingly, facilities are using light domes, which are roof-mounted domes that collect daylight and distribute it to the work spaces through polished metal shafts.

Windows and Glazing

Even in facilities with modest amounts of glazing, leaky windows lacking thermally broken sashes can be costly factors in the overall operating costs of a building. See Fig. 5.4 for a summary of strategies to improve glazing performance.

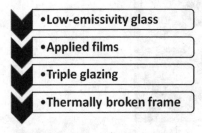

Figure 5.4   Glazing energy performance strategies.

LOW-EMISSIVITY GLASS

Low-emissivity (Low-E) glass coatings work by reflecting or absorbing infrared radiation (or heat energy). The thickness of the Low-E coating and which pane of glass is coated determine whether the window will absorb or reflect the heat energy. When installed on an interior pane of an insulated glass unit, the Low-E coating will reflect heat energy from inside the room to help reduce energy loss during winter months, thereby reducing heating costs.

When installed on the inside of the outer pane surface of an insulated glass unit, the Low-E coating will reflect or absorb heat from the outside, resulting in reduced solar gains and cooling costs during the summer.

Depending on the type of glazing used in the building, Low-E glass may be retrofitted into existing windows by replacing the entire glass panel. Low-E glazing may not be achieved with aftermarket films or spray applications (see applied films below for the attributes of this type of system). If the glazing panel in a particular window or fixed glazing unit cannot be replaced, then the entire unit will require replacement to upgrade to Low-E glazing.

TRIPLE GLAZING

Double-glazed insulated glass units have become the standard for commercial and industrial window units. A more energy-efficient option available when purchasing new window units is triple glazing, consisting of three individual panes of glass separated by two air spaces. Manufacturers claim triple glazing is 60 percent more efficient than standard double-glazed insulated units, and negates the cost of Low-E, argon, or other gas-filled double-glazed units. At a cost premium of approximately 5–10 percent more than standard double-glazed units, triple glazing is competitive with the cost of special coatings and offers equivalent performance characteristics.

APPLIED FILMS

Films applied to the interior or exterior of window glass can help to reduce solar heat gain, resulting in less heat load that must be overcome by the air conditioning system and lower bills. Films applied to older, less energy-efficient skylights can also result in significant savings without sacrificing much of the daylighting benefit of skylights. 3M, a national manufacturer of building solar films, estimates that films can remove up to 99 percent of ultraviolet rays responsible for fading and sun damage to furnishings and finishes. They also estimate that films can reduce building solar heat gain by up to 78 percent, greatly reducing the cooling load in a facility.

THERMALLY BROKEN GLAZING

All modern windows and fixed glazing have a thermally broken frame. This means that the frame construction provides a rubber or butyl separation between the interior and exterior sections of the frame to block direct conduction transfer through the frame. Older industrial facilities may contain windows or skylights without this feature. Achieving this efficiency means replacing the glazing unit in its entirety. Although there are products available that provide exterior "storm sashes" or interior glazing to provide additional efficiency, they are not well suited to industrial environments and tend to become a maintenance headache in a short time. Fortunately, replacing old windows and skylights with more efficient units can have a relatively quick payback time. The ROI varies with the degree of thermal conditioning in the space, but typically ranges from 3 to 5 years.

OPERABLE UNITS

There is considerable debate over the wisdom of allowing building users to open windows on favorable days to take advantage of outside cooling or warming breezes. Office buildings with large footprints have heating systems designed to accommodate the varied needs of users within a large space, and bring in outside air directly. The different loads within such a space, and the havoc that could result from allowing users to manipulate those loads by opening exterior windows, obviously precludes allowing them the use of operable windows. Industrial and small commercial buildings do not suffer from this need, however. In smaller office areas, it is acceptable to allow building users the ability to shut off the mechanical system and use natural ventilation and cooling in the building when they sense that outside conditions are desirable.

Building Insulation

WALL INSULATION

Industrial buildings, particularly older models, are notorious for their lack of wall insulation. The dual rationale that industrial processes generated sufficient (or excessive) heat for workers and that insulating the walls for the purpose of retaining heat would be a needless expense has been repudiated.

ROOF INSULATION

Industrial buildings, with their expansive roof areas, offer a tailor-made opportunity to save money through reducing heat loss (or gain) through the roof. Unlike office buildings, industrial and light commercial facilities can tolerate wider temperature fluctuations within the workspace. For this reason, the savings available in boosting the roof insulation R-value in industrial facilities vary with the type of facility and the manufacturing processes it houses.

OVERHEAD DOOR INSULATION

Overhead sectional doors are available with polystyrene or polyurethane cores, and insulation values ranging up to R-17.5 (Overhead Door Co. Thermacore). Uninsulated doors are common in existing industrial facilities and warehouses. The relatively low cost of replacing overhead doors, and the large amount of exterior wall area involved, can make replacement of such doors a very effective way to reduce energy consumption.

PERIMETER INSULATION

A substantial amount of heat is lost through the perimeter edge of the building, escaping at the junction of the horizontal slab and vertical foundation wall. Perimeter insulation is recommended by most building codes for any building, and is required in areas with over 2500 heating degree days. Perimeter insulation can be installed in either a vertical or horizontal manner directly under the slab. Since heat travels through the ground in concentric circles, the purpose of the perimeter in insulation is to block the transfer of heat through the short path at the slab edge. Only a minimum depth is required, since heat transfer below the insulation will be minimal. The colder the climate, the greater the thickness and depth of insulation required. For most areas of the United States, a minimum depth of 24 inches (61 centimeters), with an R-value of R-4 to R-5, is sufficient to meet the code. Regardless of code requirements, the installation of perimeter insulation for new construction is a modest investment that pays heavy dividends over the life of the building. The ROI for perimeter insulation is estimated at approximately two years.

Air Infiltration

Infiltration is the unintentional or accidental introduction of outside air into a building, typically through cracks in the building envelope and through use of doors for passage. Infiltration is sometimes called air leakage. The leakage of room air out of a building, intentionally or not, is called exfiltration. Infiltration is caused by wind, building pressurization, and buoyancy forces known commonly as the stack effect.

Because infiltration is uncontrolled, and admits unconditioned air, it is generally considered undesirable except for ventilation air purposes. Typically, infiltration is minimized to reduce dust, to increase thermal comfort, and to decrease energy consumption. For all buildings, infiltration can be reduced via sealing cracks in a building's envelope, and for new construction or major renovations, by installing continuous air retarders. In buildings where forced ventilation is provided, their HVAC designers typically choose to slightly pressurize the buildings by admitting more outside air than the amount that is exhausted so that infiltration is dramatically reduced.

Unmanaged air infiltration can have the following effects on the interior building environment:

- Places additional load on the HVAC systems to condition the additional air.
- Does not result in a filtered or dehumidified interior environment.
- Can bring in moisture and water potentially causing mold and other damage.

- Reduces the comfort of the interior space.
- Causes drafts.
- Carries dirt and pollutants.
- Can penetrate deep into the building space through ceiling spaces, wall cavities, and other unintended plenums.
- Requires additional resources for cleaning and maintenance.
- Increases the use of HVAC systems, driving up equipment and maintenance costs and the cost of utilities.
- Creates interior condensation, and the potential to increase mold problems as a result.
- Causes discomfort and is distracting to the building occupants.

Following are some solutions for eliminating facility air infiltration:

- Perform tests to determine the volume of air that is infiltrating.
- Identify location of air leaks and how to plug them.
- Go after the major leaks first. Evaluate the value of plugging the secondary leaks.
- Eliminate major sources of infiltration in existing buildings, which may entail the replacement of windows and facade systems, or modifications to and HVAC and other mechanical systems.
- Perform a cost/benefit analysis for the cost and inconvenience to the building occupants of eliminating air infiltration. In some cases, the available savings from sealing infiltration points may not warrant the expense and inconvenience.
- As part of the air infiltration cost/benefit analysis, include potential utility and maintenance cost savings.

New buildings have the potential to address this concern in the most proactive, efficient manner since no design or construction mistakes have been made yet. Following is a list of some steps that can be taken in new buildings to reduce air infiltration:

- Careful design and detailing of the building envelope, particularly between facade systems and transitions to the roof and foundation, is necessary.
- The air barrier needs to be uninterrupted. All penetrations need to be sealed and all joints in the barrier need to be properly overlapped, taped, and sealed per manufacturer's recommendations.
- The air barrier needs to be properly sealed to door and window frames.
- Insulation is not an air barrier due to where and how it is installed. Air will move through batt and loose fill insulation, and the location of foam insulation usually does not seal off the places in a wall that will allow air to pass through.
- Vents need to be sealed to the air barrier and any holes or joints in vent ducting need to be sealed to prevent leaks. Exhaust vents, stove vents, and dryer vents need dampers to prevent air from entering when the exhaust fan is not running.
- Glazing system design, assembly, and installation need to contain and prevent air from moving through to the interior.

- HVAC systems need to be designed so that they do not draw air into the conditioned space. Air for combustion for furnaces and fireplaces or woodstoves needs to be drawn in directly from the outside. Air plenums need to be sealed from outside air.
- Take care in the design process to avoid building in unintended "plenums" that outside air can follow into the building.
- The building may need an air exchange system to provide proper ventilation.

Unless a major renovation is being done, and all the occupants are moved out, it may be physically impossible to perform air infiltration building envelope improvements in an existing building.[1]

Large commercial buildings with multiple floors and high ceilings, or situated on high elevations, can experience significant air infiltration due to stack and wind effects. Stack pressure can triple for every additional floor in a structure, which amplifies the suction of air from the bottom to the top of the building.

Infiltration in commercial buildings can cause:

- Reduced thermal comfort
- Interference with the operation of building HVAC systems
- Degraded indoor air quality
- Moisture damage of the building envelope components
- Increased energy consumption

Blower door air infiltration testing has contributed significantly to understanding and measuring air leakage in and out of buildings. Prior to this technology, most air leakage was thought to occur toward the mid-height of the conditioned building envelope, primarily through doors and windows. Although these components do contribute to air leakage, air flow around these areas represents a relatively small percentage of heat loss in most dwellings. Their effects tend to be amplified by annoying high-velocity currents through relatively small gaps and cracks. Although fiberglass insulation is effective at blocking the flow of heat through solid materials (conductive heat loss), it does not work as well at blocking flow of heat through fluid substances—air or water (convective heat loss). This is why porous insulation works at its peak efficiency when air flow (pressure) through and around the material is minimized.

A mere 1/4-inch (6.4-millimeter) gap around the unsealed sides and top of a 10 foot × 10 foot (3 meter × 3 meter) sectional door is the equivalent of a 90-square inch (581-square centimeter) hole in the door, permitting 328 million cubic feet (9288 cubic decameters) of air infiltration per year.

Continuous air barrier systems can reduce air infiltration by more than 60 percent and energy consumption by up to 36 percent. A host of products have been developed to address specific air infiltration issues. One example is the Vapor Box™, a shallow plastic pan designed by a Canadian company to fit around and seal off electrical boxes.

Toilets

Appropriate technologies include high-efficiency toilets requiring not more than 1.3 gallons (4.9 liters) per flush and urinals that flush with 1 gallon (3.8 liters) or less (no automatically timed flushing systems), as well as self-closing faucets with flows of 0.5 gallons (1.9 liters) per minute for hand washing. If available, and where codes and health departments permit, nonpotable water may be used for flushing, and waterless urinals should be evaluated for suitability.

Coolers and Freezers

The Energy Policy and Conservation Act of 1975 (EPCA) established an energy conservation program for certain types of commercial and industrial equipment. Title III of the EPCA sets forth a variety of provisions designed to improve energy efficiency. The EPCA was amended by the Energy Independence and Security Act (EISA), Public Law 110-140. In particular, section 312(a) of EISA amends section 340, which defines the requirements for walk-in coolers and freezers. All new freezers and coolers manufactured or constructed in the United States are required to comply with these standards. Existing freezers and coolers are not required to upgrade to these standards, although the provisions contained in them are a good starting point in lowering energy costs for facilities containing this type of equipment.

FREEZER INFILTRATION

The EPCA requires that walk-in coolers or walk-in freezers have a method of minimizing infiltration when doors are open, regardless of the size of the door. Acceptable methods include strip curtains, spring hinged doors (also known as traffic doors or impact doors), and air curtains.

DOOR INSULATION

The EPCA requires that walk-in coolers and walk-in freezers contain door insulation of at least R-25 for coolers and R-32 for freezers.

DOOR GLAZING

The EPCA does not contain requirements that apply to the glazed portions of freezer or cooler doors. However, section 342(f)(3) requires that transparent reach-in doors for walk-in freezers and windows in walk-in freezer doors be made of triple-pane glass with either heat-reflective treated glass or gas fill. Additionally, section 342(f)(3)(B) requires that transparent reach-in doors for walk-in coolers and windows in walk-in cooler doors be either double-pane glass with heat-reflective treated glass and gas fill or triple-pane glass with either heat-reflective treated glass or gas fill.

FAN MOTORS

Evaporator fan motors of under 1 horsepower and less than 460 volts must be electronically commutated motors (brushless direct current motors) or 3-phase motors; condenser fan motors of under 1 horsepower must be electronically commutated motors, permanent split capacitor-type motors, or 3-phase motors.

INTERIOR LIGHTING

Interior light sources must have an efficacy of 40 lumens per watt or more, including ballast losses (if any). Light sources with an efficacy of 40 lumens per watt or less, including ballast losses (if any), are allowed if there is a timer or other device that turns off the lights within 15 minutes of when the walk-in cooler or walk-in freezer is not occupied by people.

COOLER AND FREEZER EFFICIENCY STRATEGIES

- Doors should be installed on open freezers and refrigerators. It may be more cost-effective to replace old refrigeration units with energy-efficient new ones.
- Night blinds should be installed on all open-faced cooling cabinets. For displays with goods that are accessed infrequently, facility managers should install day covers or plastic strip curtains to reduce energy consumption.
- Heavy plastic vertical curtains outside the walk-in cooler or freezer keep the cold air in and the warm air out.
- Energy-efficient central compressors, properly sized to match the load, can be a worthwhile investment as compressors are one of the largest energy users.
- Compressor and evaporator fan controllers for walk-in coolers and freezers, such as variable speed drives, can cut the voltage to the motor and slow down the fan when full air flow is not needed. They are most useful in units that run between 28° and 40°F (22° and 4°C) with evaporator fans that run at full speed all the time. Different models include basic units that sense when the refrigerant has ceased flowing through the evaporator coil, mid-range units that monitor data over time and activate warning lights, and top-end units that have a modem for remote or full-time monitoring. With investments as low as $100 per unit, savings can vary from 10 to 60 percent of overall refrigeration energy consumption and paybacks as low as one year.
- Remote condensers allow for the rejection of heat to the building's exterior, instead of into the retail space when the air requires cooling.
- Replace automatic defrost timers with demand-defrost controls, which run defrost cycles only when they are needed.
- Dew point controls on display cases prevent the buildup of fog on glass surfaces and the buildup of moisture on metal surfaces.
- Larger heat exchangers are more efficient than multiple, smaller units. So when renovating, try to group cabinets together to better facilitate heat removal or recovery.
- Fiber-optic lighting piped into cabinets minimizes heat input from traditional lighting.

- Lighting occupancy sensors for walk-in coolers or freezers will ensure that lights are only on when needed and will make it easier for employees to carry food in and out.
- Insulation in coolers and freezers should be inspected and upgraded regularly.
- A calculator to help determine if a company should invest in new walk-in cooler controllers is provided by the Los Angeles Department of Water and Power at http://www.ladwp.com/ladwp/cms/ladwp001511.jsp.

Space and Capacity Utilization

Capacity planning is the process of determining how much production capacity is needed by an organization to meet changing demands for its products. In the context of capacity planning, "capacity" is the maximum amount of work that an organization is capable of completing in a given period of time. The phrase is also used in business computing as a synonym for Capacity Management.

A discrepancy between the capacity of an organization and the demands of its customers results in inefficiency, either in underutilized resources or unfulfilled customers. The goal of capacity planning is to minimize this discrepancy. Demand for an organization's capacity varies based on changes in production output, such as increasing or decreasing the production quantity of an existing product, or producing new products. Better utilization of existing capacity can be accomplished through improvements in overall equipment effectiveness (OEE). Capacity can be increased through introducing new techniques, equipment, and materials; increasing the number of workers or machines; increasing the number of shifts; or acquiring additional production facilities.

Capacity is calculated as: (number of machines or workers) × (number of shifts) × (utilization) × (efficiency).

The three main categories of capacity planning are lead strategy, lag strategy, and match strategy. They are defined as follows:

- Lead strategy is adding capacity in anticipation of an increase in demand. Lead strategy is an aggressive strategy with the goal of luring customers away from the company's competitors. The possible disadvantage to this strategy is that it often results in excess inventory, which is costly and often wasteful.
- Lag strategy refers to adding capacity only after the organization is running at full capacity or beyond due to increase in demand. This is a more conservative strategy. It decreases the risk of waste, but it may result in the loss of possible customers.
- Match strategy means adding capacity in small amounts in response to the changing demand of the market. This is considered to be a middle, or moderate strategy.

Capacity planning consists of long-term planning that determines a company's overall level of resources and readiness to make changes in their business. In the context of systems engineering, capacity planning is used during system design and system performance monitoring. It extends over a timeline that is long enough to allow the managers to obtain resources they need to implement the strategy. Capacity decisions affect the production lead time, customer responsiveness, operating costs, and the company's overall ability to compete. Inadequate capacity planning can lead to loss of customers and business. Excess capacity can drain the company's resources and prevent investments into more lucrative and profitable areas.

INTEGRATED DESIGN PRINCIPLES

Integrated design   Integrated design involves the use of a collaborative, integrated planning and design process that initiates and maintains an integrated project team in all stages of a project's planning and delivery; establishes performance goals for siting, energy, water, materials, and indoor environmental quality along with other comprehensive design goals; ensures incorporation of these goals throughout the design and life cycle of the building; and considers all stages of the building's life cycle, including deconstruction.

Commissioning   Total building commissioning involves practices tailored to the size and complexity of the building and its system components, in order to verify performance of building components and systems and help ensure that design requirements are met. This should include a designated commissioning authority, inclusion of commissioning requirements in construction documents, a commissioning plan, verification of the installation and performance of systems to be commissioned, and a commissioning report.

Optimize Energy Performance

ENERGY EFFICIENCY

Establish a whole building performance target that takes into account the intended use, occupancy, operations, plug loads, other energy demands, and design to earn the ENERGY STAR targets for new construction and major renovation where applicable. For new construction, reduce the energy cost budget by 30 percent compared to the baseline building performance rating per the American Society of Heating, Refrigerating and Air-Conditioning Engineers, Inc. (ASHRAE) and the Illuminating Engineering Society of North America (IESNA) Standard 90.1-2004, Energy Standard for Buildings. For major renovations, reduce the energy cost budget by 20 percent below the prerenovation 2003 baseline. It should be noted that the U.S. Department of Energy (DOE) requires compliance with the 2009 International Energy Conservation Code (IECC) and ASHRAE 90.1-2007 to qualify for any applicable credit under the American Recovery and Reinvestment Act (ARRA) of 2009 and to meet the nationwide energy-efficiency goals.[2]

MEASUREMENT AND VERIFICATION

In accordance with DOE guidelines issued under section 103 of the Energy Policy Act of 2005 (EPAct), install building level utility meters in new major construction and renovation projects to track and continuously optimize performance. Compare actual performance data from the first year of operation with the energy design target. After one year of occupancy, measure all new major installations using the ENERGY STAR Benchmarking Tool for building and space types covered by ENERGY STAR. Enter data and lessons learned from sustainable buildings into the High Performance Buildings Database, www.eere.energy.gov/femp/highperformance/index.cfm.

PROTECT AND CONSERVE WATER

Indoor water   Employ strategies that in aggregate use a minimum of 20 percent less potable water than the indoor water use baseline calculated for the building, after meeting the Energy Policy Act of 1992 fixture performance requirements.

Outdoor water   Use water-efficient landscape and irrigation strategies, including water reuse and recycling, to reduce outdoor potable water consumption by a minimum of 50 percent over that consumed by conventional means (plant species and plant densities). Employ design and construction strategies that reduce storm-water runoff and polluted site-water runoff.

ENHANCE INDOOR ENVIRONMENTAL QUALITY

Ventilation and thermal comfort   Meet the current ASHRAE Standard 55-2004, Thermal Environmental Conditions for Human Occupancy, including continuous humidity control within established ranges per climate zone, and ASHRAE Standard 62.1-2004, Ventilation for Acceptable Indoor Air Quality.

Moisture control   Establish and implement a moisture control strategy for controlling moisture flows and condensation to prevent building damage and mold contamination.

Daylighting   Achieve a minimum of daylight factor of 2 percent (excluding all direct sunlight penetration) in 75 percent of all space occupied for critical visual tasks. Provide automatic dimming controls or accessible manual lighting controls and appropriate glare control. See Fig. 5.5 for a photograph illustrating the benefits of warehouse daylighting.

Low-emitting materials   Specify materials and products with low pollutant emissions, including adhesives, sealants, paints, carpet systems, and furnishings.

Protect indoor air quality during construction   To protect installed HVAC systems from contamination during construction, contractors should follow the recommended approach of the Sheet Metal and Air Conditioning Contractor's National Association (SMACNA), *Indoor Air Quality Guidelines for Occupied Buildings under Construction, 1995.* After construction and prior to occupancy, building managers should conduct a minimum 72-hour flush-out with maximum outdoor air consistent with achieving relative humidity no greater than 60 percent. After occupancy, the flush-out should continue as necessary to minimize exposure to contaminants from new building materials.

Figure 5.5   Light interior colors reduce the number of required light fixtures.

REDUCE ENVIRONMENTAL IMPACT OF MATERIALS

Recycled content   For EPA-designated products, use products meeting or exceeding EPA's recycled content recommendations. For other products, use materials with recycled content such that the sum of postconsumer recycled content plus one-half of the preconsumer content constitutes at least 10 percent (based on cost) of the total value of the materials in the project.

Bio-based content   For United States Department of Agriculture (USDA)-designated products, use products meeting or exceeding USDA's bio-based content recommendations. For other products, use bio-based products made from rapidly renewable resources and certified sustainable wood products.

Construction waste   During a project's planning stage, identify local recycling and salvage operations that could process site-related waste. Program the design to recycle or salvage at least 50 percent construction, demolition, and land clearing waste, excluding soil, where markets or on-site recycling opportunities exist.

Ozone-depleting compounds   Eliminate the use of ozone-depleting compounds during and after construction where alternative environmentally preferable products are available, consistent with either the Montreal Protocol and Title VI of the Clean Air Act Amendments of 1990, or equivalent overall air quality benefits that take into account life-cycle impacts.

See Table 5.3 for a summary of sustainability performance goals.

Tip Box

TABLE 5.3   SUSTAINABILITY PERFORMANCE GOALS

Pursue energy efficiency

   20% better than ASHRAE 90.1

Measure and verify

   Utility meters

   Performance metrics

   Benchmarking tools

Conserve water

   Use 20% less indoors than baseline

   Reduce outdoor usage by 50%

Enhance indoor environmental quality

   Meet ASHRAE Standard 55-2004

   Perform air infiltration control

Daylighting

   Achieve 2% daylighting in 75% of spaces

Low-emitting materials

   Install or replace with low-VOC materials

Protect indoor air quality

   Conduct minimum 72 hour flushout of new spaces

Reduce environmental impact

   EPA products: Meet EPA recycled content recommendations

   Other products: 10% recycled content (post + 50% preconsumer)

   Bio-based content: Exceed USDA's recommendations

   Eliminate the use of ozone-depleting compounds

Construction waste

   Reuse or recycle 50% of construction waste

Endnotes

1. Kearns, Tom. The Façade Group, LLC. "The Effects of Air Infiltration in Commercial Buildings." http://www.docstoc.com/docs/15419431/The-Affects-of-Air-Infiltration-in-Commercial-Buildings. Accessed Feb 17, 2010.
2. EnergyStar.gov. "Guiding Principles for Federal Leadership in High Performance and Sustainable Buildings." http://www.energystar.gov/ia/business/Guiding_Principles.pdf Accessed March 10, 2010.

6

EQUIPMENT

Boilers and Pressure Vessels

Boilers enjoy widespread use in commercial facilities as a cost-efficient means of generating heat for space heating and process heat. See Fig. 6.1 for a photograph of a typical boiler installation.

On the sustainability front, Burnham Commercial offers a multi-pass boiler that provides more efficient operation than a conventional single- or double-pass boiler. A multi-pass boiler directs flue gases through multiple combustion chambers, extracting the maximum amount of heat out of the boiler flue gases. Single- or double-pass boilers fire directly into a combustion chamber before exhausting the flue gases upward between the boiler flueways where they exit the building through the vent. In comparison, a multi-pass boiler directs flue gases into several smaller passes, which allows the sections to absorb more heat than if the flue gases simply passed through the boiler sections once.

Burnham's system also utilizes a return water mixing tube, which distributes water evenly throughout the whole assembly. Distributing cool return water in this manner reduces areas of extreme hot and cold, eliminating undue stress on the boiler castings. See Table 6.1 for a list of savings factors for high-efficiency boilers. See Table 6.2 for an *ROI Quik-Calc* to estimate boiler conversion savings.

Chillers

A clear understanding of two measures of chiller efficiency—the design-efficiency rating and the Nonstandard Part Load Value (NPLV) rating—can help organizations obtain the best capital cost and energy efficiency when acquiring new chillers. It also may help facility managers understand why they may not be getting the level of energy efficiency they expect from existing chillers.

Figure 6.1   **High-efficiency boiler operation.**

Tip Box

TABLE 6.1   HIGH-EFFICIENCY BOILERS

Savings factors in using high-efficiency boilers

1. Condensing boilers require a low return water temperature to operate at their highest efficiency.

2. Condensing boilers can function with smaller venting pipe (stainless steel is required for larger boilers).

3. Smaller systems can use PVC venting pipe, which can be vented directly through sidewalls.

4. Heating coils and radiators should be sized for a higher rate of heat transfer at lower supply water temperatures.

5. Systems should be designed with lower flow rates. This means that piping, pumps, and valves should be smaller than those used in mid-efficiency boilers.

Many system designers mistakenly believe that "more is better," and a larger chiller will provide more efficient operation. As a result, they use both the design-efficiency and the NPLV ratings in the specifications they provide to facility managers and building owners. This approach is flawed, as it is based on a couple of misconceptions.

The first misconception is that the NPLV rating only measures chiller performance at off-design conditions. In fact, the NPLV rating already includes measures for both the design efficiency and off-design efficiency of the chiller.

ROI Quik-Calc

TABLE 6.2   HIGH-EFFICIENCY BOILER SAVINGS ROI
Standard efficiency:
80% efficient/on-off boilers/2 boilers
Gas usage (GJ): 4800
Boiler cost: $50,000
High efficiency:
88% efficient/condensing boilers/4 boilers
Gas usage (GJ): 5500
Boiler cost: $110,000
Estimated savings of high-efficiency boilers per year: $6400
Simple payback period: 9.4 years

Off design performance is critical, because most chillers run most often under partial load or at off-design temperatures. Specifically, 99 percent of chiller operating hours are spent during off-design conditions.

When energy codes or utility rebates require inclusion of the design-efficiency rating in the specification, it is better for the designer to specify the maximum kilowatts (kW) or kilowatts per tonnage of refrigeration (kW/TR) required by the code or rebate. This is because a lower value could result in higher capital costs with no reduction in annual energy costs. This disparity is leading more code-writing agencies to recognize the NPLV rating.

Chiller-efficiency specifications that specify both the NPLV rating and the design-efficiency rating may hinder the designer's ability to meet the owner's goals, if the objective of the specification is to attain the lowest capital cost for similar annual energy. Using two ratings can create inequalities in annual energy-consumption comparisons, which also result in higher capital costs passed on to the owner. Also, the design-efficiency rating usually has little practical impact on electrical-demand charges and wiring size.

Instead of using both ratings, the best chiller-efficiency specification uses the NPLV rating by itself. For power-wire sizing, specifying the maximum full-load amperage and the minimum power factor eliminates all ambiguity about actual size requirements. If energy codes or utility rebates require that the specification include the design-efficiency rating, the maximum allowable kW or kW/TR should be specified.[1]

A second misconception is that a chiller with good efficiency at design conditions will automatically have a good NPLV rating. In fact, chillers can have the same design efficiency but have NPLV ratings that vary widely, depending on capital cost. That's because chillers can have different off-design efficiencies. See Table 6.3 for an *ROI Quik-Calc* for estimating HVAC savings from converting to a more efficient system.

ROI Quik-Calc

TABLE 6.3   FEDERAL ENERGY MANAGEMENT PROGRAM CALCULATORS

These Federal Energy Management Program energy cost calculators compare the energy costs for selected units with the energy cost for baseline unit for various energy costs, efficiency levels, size variations, and hours of operation for the following types of equipment:

- Air-chilled coolers
- Boilers
- Commercial unitary air conditioners
- Commercial heat pumps
- Water-cooled chillers

Source: U.S. Department of Energy
Federal Energy Management Program
Energy Efficiency and Renewable Energy
http://www1.eere.energy.gov/calculators/buildings.html

Motor-Driven Equipment

Advances in motor design, motor manufacture, and legislation requirements are the most important reasons for higher efficiencies in newer motors. Load levels below 50 percent will typically show lower efficiencies. The current level increases with lower voltage levels, which tends to decrease the efficiency. Commonly, a motor running at rated voltage, or only slight over, will show the highest efficiency. Voltage unbalance increases the average current slightly, but the losses in one or more of the phases are substantial. This causes the efficiency to drop drastically. Current level increases strongly with mild increases in voltage distortion, equating higher currents to lower efficiencies. Broken rotor bars or end rings cause the motor to need more slip for delivery of the required torque. This has the effect of additional current in the stator, which also decreases the operating efficiency. Motors operating at low efficiencies burn up larger amounts of heat in losses. This does not automatically mean that these motors will run hotter. Older motor designs, for example, include larger fans for additional heat reduction. Unfortunately, this causes additional losses from friction and windage created by these larger fans. Newer motor designs frequently allow the motor to operate under higher temperatures, since the quality of the insulation material has improved with respect to older motors. This allowed the motor designers to additionally increase the efficiency of the motors by decreasing the amount of cooling and windage loss with a smaller fan.

For any motor, however, any reason causing it to operate at lower efficiencies is reason for expecting higher operating temperatures, and lower life due to insulation deterioration.

Low loads are sources of poor power factor, and low efficiencies. Unless there are clear reasons for doing otherwise, loading of 75–100 percent is viewed as optimal for efficiency.

One way to carry out this necessity is to determine the efficiency of the motor and correct issues with the surrounding system that directly relate to, or affect, the motor's efficiency. Correcting issues, such as cavitating pumps, overcurrent situations, and improperly tuned drives, can increase the efficiency and life of the motor while reducing the maintenance costs, downtimes, and production losses.

To illustrate the potential of motor efficiency savings, consider a 200-horsepower motor running at 8000 hours per year at 85 percent efficiency. This motor would:

1 Cost the facility $105,318 per year to run (not including penalties for poor power factor and any carbon tax paid).
2 Cause the facility to use 1,053,176 kilowatt hours (kWh) per year on this motor.
3 Cause approximately 790 tons (717 tonnes) of carbon dioxide (CO_2) to be produced and emitted into the atmosphere.

Swapping this 85 percent efficient motor for a National Electrical Manufacturers Association (NEMA) Premium 96.2 percent efficient motor will result in the following savings:

1 An annual energy savings of $12,262, which by the end of the first year more than pays for the motor.
2 An annual kWh reduction of 122,615, thereby reducing energy demands.
3 A decrease in CO_2 buildup by roughly 91 tons (83 tonnes).

Due to utility rebates currently in existence and the possibility of a government-introduced rebate program, the payback period for this motor is approximately 8.6 months.[2]

Spray Booths

Parameters that affect transfer efficiency and paint coverage are as follows:

- Equipment setup
- Solvent and solids content
- Spray coating equipment
- Shape of workpiece
- Changing conditions
- Operator training
- Paint booth

PAINT BOOTHS

Why is a paint booth important? It is important from worker exposure, quality, and paint usage perspectives. A paint booth is intended to collect overspray paint and to remove solvent fumes from the work area. If painting is done in an area with no ventilation, fumes will build, resulting in a fire and health hazard. Also, oversprayed paint will fall back onto newly painted surfaces, causing quality problems. Paint booths also eliminate drafts and odd currents of air that could otherwise carry paint away from the part and onto neighboring cars and buildings.

It is important to provide makeup air for the booth. If a booth exhausts 3000 cubic feet per minute (85 cubic meters per minute), then an equal amount of air must be brought into the building to avoid negative pressure situations. Insufficient makeup air will result in reduced flow through the booth, eliminating booth benefits. It should also be remembered that paint booths and spray booths must comply with the local building, fire, and mechanical codes such as the International Building Code (IBC), the International Fire Code (IFC), and the International Mechanical Code (IMC).

OPERATOR TRAINING

Spray painting personnel have a tremendous effect on transfer efficiency and coverage. If a painter does everything correctly, the transfer efficiency will be no better than that quoted in the paint equipment sales catalog. There are some common waste problems that can be partially solved by training. But the operator should not bear the blame for all problems resulting in waste. Often painters have not been properly trained or do not have the correct tools such as measuring equipment or high-efficiency spray equipment. Following are some tips for improving efficiency in commercial painting operations:

- **Clear the gun:** Triggering the paint gun when the gun is pointed at the floor or ceiling or anywhere except the part being painted wastes paint.
- **Move the gun in an arc:** Move the paint gun in an arc so that ends of the stroke are too far away and too close in the middle. Paint guns should be moved parallel to the surface.
- **Using too much overlap:** A 50 percent overlap pattern is usually recommended to avoid heavy and light areas.
- **Applying an extra thick coating:** A paint manufacturer may recommend 1.5-mil (38-micron) thicknesses, but painters may apply 3 mils (76 microns) in actual practice. The result is twice as much paint being used.
- **Using an incorrect fan pattern:** A wide fan pattern is great for wide open spaces. It is not so good for painting narrow edges, because a narrow edge will occupy only a small portion of the fan. The rest of the paint is overspray.
- **Needlessly increasing pressure:** Increasing fluid and atomizing pressure well above the recommended setting is common. The result is usually more paint being used.
- **Holding the gun at an angle:** Spray guns should be pointed perpendicular to a surface. Holding it at an angle results in some or the entire spray pattern being too far away from the surface.

SUMMARY

There is no one overall solution that will result in minimum paint usage. Efficient application equipment, equipment setup, using high-solids and water-based paints, and employee training will help. It is also important to consider the performance of the paint, and to educate purchasing personnel that low paint cost per gallon or the price of a paint gun may only be part of the equation.

Ovens

ENERGY-EFFICIENCY STRATEGIES

Tune the burner   Burner tuning sessions typically require less than a day per unit and may return the cost of the tuning in less than a year. (1–3% of fuel use)

Reduce air infiltration   Air movement through a process oven or furnace and even a large field-erected boiler can make for significant fuel loss. Tools such as infrared analysis can be used to identify where heat is being lost through a boiler, furnace, or oven. This is often a matter of closing up all openings with insulation blankets and making minor repairs at the leaking areas. (2–5% of fuel use)

Linkage minus controls   Linkages allow limited tuning capability of combustion equipment, limiting the ability to optimize the fuel/air ratio. The answer is a change to the control strategy where the oven has an actuator installed for both the gas and the air. This allows for more precise control of fuel/air ratios. (1–3% of fuel use)

Install oxygen trim controls   The excess air in the stack can be measured and metered for a particular firing rate by adding a probe to the stack. This probe can transmit instructions to the fuel air controls to optimize the air flows. This technique requires a stack damper and separate fuel and air controls.

Pulse firing controls   Process oven or furnace burners operate more efficiently when they can run at high fire in an optimized condition. Instead of running at some turndown condition, this strategy allows the burners to operate at high fire but in a pulsed condition to get the turndown required. Implementing this strategy requires specially designed fuel trains and high-cycle rate fuel control valves.

Install more efficient burners   The efficiency of the oven may be limited by the fundamental type of burner installed in the oven. A fixed excess air style of burner in an older oven, for instance, will consume more energy than a variable burner. (2–5% savings)

Install heat recovery capability   There may be an opportunity to use heat from the flue gases elsewhere in the production process. These uses may include preheating combustion air, feed water, and/or even to preheat materials that are about to head into an oven or furnace. (3–6% savings)

Check dryers/oven controls   Issues affecting energy consumption of dryers and ovens:

■ How is the drying cycle set?
■ Can the drying cycle be reduced?
■ How are the dryers arranged and controlled?
■ Are they controlling solely by dry bulb temperature?
■ Does the control algorithm consider outside ventilation air humidity?
■ Have relative humidity sensors been installed on each unit?
■ What is the temperature set point and what is it based on?

Evaluate oven heat distribution   Poor interior oven heat distribution may be due to various factors. These factors include blocked ducts, inoperable recirculation fans, and poor burner adjustments. Have ovens professionally evaluated at least once per year to ensure they are operating to specification and not wasting fuel and creating an excessive amount of spoiled product. (2–5% savings)

Check thermal oxidizer and oven temperature settings   Check the temperature recorder from the oven's thermal oxidizer. If it has a deep saw-tooth pattern, the oven's digital temperature controller may require tuning since it is allowing the oven to burn hot and waste fuel. (2–4% savings)

Consider oven pressure controls   Ovens and furnaces with features such as combined flues, tall flues, and exhaust fans can draft room air from undesirable spaces. Draft or suction in the furnace or oven needs to be precisely controlled for optimizing energy efficiency. This can be accomplished with pressure controls and stack dampers. (3–6% energy savings)

Control ventilation air   Ventilation air is required to remove moisture and/or contaminants in an oven or furnace. Most industrial ovens must allow the lower explosive limit (LEL) in the oven to reach 25 percent. However, any level higher than this is pure waste. Measure the air flows and adjust to optimize this issue. (4–8% energy savings)

Check fuel train vent valves   Valves in double block and bleed fuel train systems sometimes leak through when the unit is firing. This allows fuel to vent to the exterior, resulting in an expensive loss over time.

Check oven/door seals   Use visual surveys or ultrasonic equipment to check for deteriorated door seals. Add this check to the periodic preventative maintenance review. (1–2% energy savings)

Consider oven insulation/refractory upgrades   Oven skin temperatures over 160°F (71°C) represent a potential burn hazard and indicate heavy heat loss from the oven. Consider installing a refractory or insulation upgrade. (1–3% energy savings)

Keep heat transfer surfaces clean   Dust and soot build up over time. Heat transfer efficiency declines even with 1/16 inch (1.6 millimeters) of soot or dust coating. Add oven cleaning to the periodic list of preventative maintenance. (2–6% energy savings)

Carefully manage loads   Small loads, heavy baskets and dunnage, and long loading cycle times where furnaces or ovens are sitting idle make for considerable waste. Minimize this intermediate waste time in the middle of the shift. In some cases programmable logic controller (PLC) codes can be modified to change the operating parameters, resulting in 2–5% energy savings.

Evaluate weekend shutdown/startup cycles   Not running equipment is the best form of economy. Time the warm-ups to complete just prior to the start of production or the period when the heat is needed. Reduce the amount of time the oven is sitting idle after reaching temperature.

Review thermal oxidizer air flows   Review the amount of air being sent to the thermal oxidizer. Check the air flows to make sure they are optimal. The load in a thermal oxidizer is directly proportional to the amount of air being sent to it. (2–6% energy savings)

Consider a technology change   A regenerative unit installed in place of a thermal oxidizer or afterburner swaps heat back and forth to a recovery chamber and may cut fuel consumption by 50 percent. Replacing an old fire tube or water tube type boiler with new steam generators can trim boiler fuel costs by 15–20 percent. (10–30% energy savings)[3]

Conveying Equipment

A conveyor system is a common piece of mechanical handling equipment that moves materials from one location to another. Conveyors are especially useful in applications involving the transportation of heavy or bulky materials. Conveyor systems allow quick and efficient transportation for a wide variety of materials, which makes them very popular in the material handling and packaging industries. Many kinds of conveying systems are available, and are used according to the various needs of different industries. See Fig. 6.2 for a photograph of a typical conveyor setup in a distribution center.

TYPES OF CONVEYOR SYSTEMS

- Belt-driven roller conveyor for cartons and totes
- Gravity roller conveyor
- Gravity skate wheel conveyor
- Belt conveyor

Figure 6.2   Typical conveyor setup in a distribution facility.

- Wire mesh
- Plastic belt
- Belt-driven live roller
- Line shaft roller conveyor
- Chain conveyor
- Screw conveyor
- Chain-driven live roller conveyor
- Overhead conveyor

Conveyor systems have not been at the forefront of energy savings or sustainability, but their widespread use in manufacturing and distribution make them ideal candidates for innovation. Some companies are taking up the mantle. For instance, Hytrol has introduced their E24 convey or and E24EZ conveyor systems, which they claim are more sustainable than the competing conventional systems. Hytrol claims the E24 offers energy savings as high as 60 percent over conventional conveyors, and its 125,000-hour life expectancy is approximately six times longer than that of conventional roller systems. In another example, DynaCon conveyors are considered eco-friendly because of the ability to repurpose them and replace damaged components when needed. Accessories can be easily added at any time, and the DynaCon's plastic belt conveyors are ISO Class 3 clean-room ready. DynaCon states their conveyor's adaptability means it should never be obsolete.[4]

Vacuum Equipment

Compressed air vacuum generators are common in industry. Palletizers, material-handling systems, pick-and-place operations, drum-type vacuum cleaners, and packaging applications are just a few examples. Each generator is mounted in close proximity to the point of use, with supply tubing connecting the vacuum device to a central compressed air system.

Vacuum generators are considered to be reliable, lightweight, compact, and quiet. They have no moving parts and can be mounted directly onto production machinery. Their maintenance requirements are minimal, and they are produced in aluminum, plastic, and corrosion-resistant housings for harsh applications. Replacement or repair is relatively simple and requires no special tools or training.

Given these attributes, there would appear to be little reason to consider an alternative system. There are reasons, however, and they involve efficiency and energy savings. What at first appears to be a winning way to produce vacuum turns out to be a technology with inadequate performance. Conventional vacuum systems rely on compressed air systems, consume large amounts of electricity, develop leaks periodically, and require consistent maintenance.

The alternative is a central facility vacuum system. Like compressed air, vacuum can be generated at a central location and distributed through a network of headers and drops. Unlike compressed air, vacuum supply piping can be made of light, flexible, inexpensive, and easy-to-install plastic.

With careful attention to sizing the pipe diameter and isolation valve placement, a central vacuum system with an electric motor-driven vacuum pump can provide the same service as an air compressor/vacuum generator system using only a fraction of the energy. If a central vacuum system is impractical, small area systems can service production equipment groups.

A central vacuum system has all the advantages of individual compressed air vacuum generators, except that there is no pump or motor noise at the point of use. Vacuum tubing in electric systems takes up about the same space as compressed air supply tubing. These systems also have no heat problems or oil mist. The servicing schedule for an electric motor-driven vacuum pump is usually identical to that of an air compressor, and a variety of electric vacuum pumps are manufactured specifically for chemical resistance.

Many applications use hundreds of compressor horsepower to generate vacuum. Replacing these systems with dedicated electric vacuum pumps can save thousands of dollars annually.[5]

Vacuum generators powered by compressed air represent an inefficient method of meeting this plant need. While sales literature may highlight the benefits of using vacuum generators, it often fails to portray the total energy consumption and maintenance picture. Even though the venture vacuum pumps themselves may be quiet, low-cost, and environmentally friendly, the large air compressors required as part of the systems are not. They are expensive, high-energy consumptive, and noisy pieces of equipment. In many applications, electric motor-driven vacuum pumps can achieve the same performance as vacuum generators while using as little as one-fourth the energy. In fact, replacing compressed air vacuum generators might be one of the best methods remaining for increasing production energy efficiency and taking overworked air compressors off-line.

Welding Equipment

Traditional welding is an old process offering little opportunity for technology improvements, meaning the industry must do all it can to improve the process of welding. Welding sustainability best practices should include:

- Ensuring all production waste is reused, recycled, or used as raw material in other processes.
- Significantly reducing hazardous welding fumes.
- Ensuring effective ventilation systems are used by all welders.
- Ensuring welding slags and other product waste is recycled or used as raw material in other processes.

FRICTION STIR WELDING

Invented in 1991 by Wayne Thomas of The Welding Institute in the United Kingdom, friction stir welding (FSW) is a joining technique that has gained attention as an attractive alternative to traditional fusion welding for many metals, but especially for joining aluminum alloys. FSW's ability to create a solid-state bond makes it a prime candidate for materials that are difficult to join using traditional arc welding techniques. Aluminum and aluminum alloys, which can be difficult to weld due to their high thermal conductivity, large freezing range, rapid formation of oxide film over the liquid weld puddle, and tendency to form porosity and solidification cracks, were the first materials to be joined using FSW. However, the technology has also been effective on copper, lead, magnesium, titanium, zinc, mild steel, selected stainless steels, and nickel alloys.

FSW significantly reduces residual stresses compared to traditional welding techniques and creates strong and ductile joints with low distortion, shrinkage, and porosity. The FSW process is relatively simple, is fully mechanized, and can function well in any position (gravity has no effect on the FSW process, as opposed to fusion welding). In addition, FSW consumes less energy than fusion welding and eliminates the need for filler wire, thereby making FSW a more environmentally friendly technique.

Grinding Equipment

An industrial grinding machine consists of a power-driven grinding wheel spinning at the required speed, and a bed with a fixture to guide and hold the workpiece. The grinding head can be controlled to travel across a fixed workpiece or the workpiece can be moved while the grinding head stays in a fixed position. Grinding machines remove material from the workpiece by abrasion, which can generate substantial amounts of heat. For this reason, grinding equipment often incorporates a coolant to cool the workpiece so that it does not overheat and exceed its tolerance. The coolant also helps to protect the machinist as the heat generated may cause burns. High-precision grinding machines, such as cylindrical and surface grinders, remove very little material and, as a result, generate much less heat.

Grinding machines come in a wide variety of types, serving many industrial needs:

- **Belt grinder.** Belt grinding is a versatile process suitable for all kind of applications, such as finishing, deburring, and stock removal.
- **Bench grinder.** A bench grinder usually has two wheels of different grain sizes for roughing and finishing operations and is secured to a workbench. It is used for shaping tool bits or various tools that need to be made or repaired. Bench grinders are operated manually.
- **Cylindrical grinder.** With a cylindrical grinder, the workpiece is rotated and fed past the wheel(s) to form a cylinder. A cylindrical grinder can have multiple grinding wheels. This type of grinder is commonly used to make precision rods.
- **Surface grinder and wash grinder.** A surface grinder has a head that is lowered against a moving workpiece on a table.
- **Tool and cutter grinder.** A tool and cutter grinder performs mainly as a drill bit grinder, but may also be used to make specialty drills and specialty implements such as micro tools.
- **Jig grinder.** A jig grinder's primary function is to grind holes and pins. It can also be used for complex surface grinding to finish work started on a mill.
- **Gear grinder.** A gear grinder is usually employed as the final machining process when manufacturing a high-precision gear. The main function of these machines is to remove the remaining few thousandths of an inch of material left by other manufacturing methods.

Gleason has introduced a new line of TITAN® Grinding Machines designed to reduce finish grinding times by as much as 50 percent on cylindrical gears up to 59 inches (1500 millimeter) in diameter.

This new process, called *Power Grind,* enables users to reduce grinding production times by as much as 50 percent, by first using threaded wheel grinding to "rough" gears much faster and then profile grinding to achieve optimal gear quality, surface finish, and complex gear modifications in the finishing operation.

Drying Equipment

Refrigerated dryers represent the largest segment of installed equipment and currently provide the greatest amount of drying capacity in the United States. It is the least expensive of the four commercial methods of drying air. Refrigerated dryers in their simplest form are large refrigerators containing compressed air tubing. The function of the refrigeration system is to reduce the temperature of the air and condense the water vapor into liquid water. This liquid water is then separated from the compressed air stream and drained.

Desiccant dryers operate on a simple principle—pass the moist compressed air over a material of very low moisture content. Doing so causes the water vapor to move from the relatively "wet" air to the relatively "dry" desiccant.

Deliquescent dryers are single-tower dryers that consume the deliquescent desiccant, usually a salt, during operation. An example is the small silica gel packets that are packed with new cameras and some electronic gear. Deliquescent dryers have advantages: they are simple, they have a low initial cost, and they require no electrical power. When properly applied, these dryers reduce or suppress the dew point of the compressed air by 20 to 40°F (−6.7 to 4.4°C). This type is not common in general plant compressed air systems.

Membrane dryers deal with the separate gaseous components of compressed air and permit the water vapor and some oxygen molecules to diffuse across a membrane. The driving force across the membrane is the differential between the pressure and relative humidity of the compressed air and that of the atmospheric air surrounding the membrane fibers. The amount of air lost in this process ranges from 25 to 40 percent of the inlet air flow. Other than a few pioneering applications, use of this approach for air drying is currently limited to air flows less than 50 standard cubic feet (1.4 cubic meters) per minute for specific point-of-use applications.

Dryer efficiency lies mostly in making the selection of the appropriate type, and matching the capability to the process need.[6] See Table 6.4 for dryer equipment energy-saving tips.

Tip Box

TABLE 6.4   DRYING EQUIPMENT ENERGY-SAVING TIPS

- Pay proper attention to the drying equipment and upstream processes (savings potential 10% of the total energy load).
- Keep heat exchanger surfaces clean.
- Implement a program of regular inspection and preventive maintenance.
- Upgrade or add monitoring and control equipment.
- Maintain proper burner adjustments and monitor flue gas combustibles and oxygen.
- Relocate combustion air intake to recover heat from other processes (or from within the building).
- Schedule production so that each furnace/kiln or drying oven operates near maximum output.
- Maintain equipment insulation.
- Replace warped, damaged, or worn furnace doors and covers.

Computers and Office Equipment

Electronic waste, e-waste, e-scrap, or Waste Electrical and Electronic Equipment (WEEE) describes loosely discarded, surplus, obsolete, or broken electrical or electronic devices. Environmental groups claim that the informal processing of electronic waste in developing countries causes serious health and pollution problems. Some electronic scrap components, such as cathode-ray tubes (CRTs), contain contaminants such as lead, beryllium, mercury, and brominated flame retardants. Activists claim that even in developed countries recycling and disposal of e-waste may involve significant risk to workers and communities, and great care must be taken to avoid unsafe exposure in recycling operations and leaching of material such as heavy metals from landfills and incinerator ashes. Scrap industry and U.S. Environmental Protection Agency (EPA) officials agree that materials should be managed with caution, but that environmental dangers of unused electronics have been exaggerated by groups that benefit from increased regulation.

The Waste Electrical and Electronic Equipment Directive (WEEE Directive) is the European Community Directive 2002/96/EC on WEEE that, together with the Restriction of Hazardous Substances (RoHS) Directive 2002/95/EC, became European Law in February 2003, setting collection, recycling, and recovery targets for all types of electrical goods.

The directive imposes the responsibility for the disposal of WEEE on the manufacturers of such equipment.

The Electronic Waste Recycling Act of 2003 (2003 Cal ALS 526) (EWRA) is a California law to reduce the use of certain hazardous substances in certain electronic products sold in the state. The act was signed into law in September 2003.

Under the California law, all old cathode ray tube (CRT), liquid crystal display (LCD), and plasma display devices contained in televisions, computers, and other electronic equipment with a screen size over 4 inches (10 centimeters) measured diagonally are covered by the act. These devices may not contain greater than the allowed concentrations of any of these four materials (by weight):

- Cadmium: 0.01 percent
- Hexavalent chromium: 0.1 percent
- Lead: 0.1 percent
- Mercury: 0.1 percent

The act also requires retailers to collect an Electronic Waste Recycling Fee from consumers who purchase covered devices.

The Resource Conservation and Recovery Act (RCRA), enacted in 1976, is the principal federal law in the United States governing the disposal of solid waste and hazardous waste.

This law covers only CRTs, though state regulations may be more restrictive. Various states have also enacted separate laws concerning battery disposal. As of 2008, 17 states have producer responsibility laws in some form. In all, 35 states have or are considering electronic waste recycling laws.

Office equipment, including photocopiers, which previously ended up in landfill sites, are no longer eligible for standard landfill disposal under the WEEE Directive. The WEEE Directive affects those involved in the manufacture, selling and distribution, recycling, and treating of any electronic equipment.

Affected by this directive are household appliances, information technology equipment of all kinds, telephone/telecommunications equipment, audiovisual gear, lighting equipment, electrical and electronic tools, hospital and medical devices and automatic dispensers, and of course, office equipment including photocopiers.

Computer recycling consists of the recycling or reuse of obsolete computers. Recycling of computers includes both finding another use for materials (such as donation to charity) and having systems dismantled in a manner that allows for the safe extraction of the constituent materials for reuse in other products.

CORPORATE COMPUTER RECYCLING

Businesses seeking a cost-effective way to recycle large amounts of computer equipment responsibly face a more complicated process. They also have the option of contacting the manufacturers and arranging recycling options. However, in cases where the computer equipment comes from a wide variety of manufacturers, it may be more efficient to hire a third-party contractor to handle the recycling arrangements.

Professional IT Asset Disposition (ITAD) firms specialize in corporate computer disposal and recycling services in compliance with local laws and regulations and also offer secure data elimination services that comply with data erasure standards. Companies that specialize in data protection and green disposal processes dispose of both data and used equipment while at the same time employing strict procedures to help improve the environment. Some companies will pick up unwanted equipment from businesses, wipe the data clean from the systems, and provide an estimate of the product's remaining value. For unwanted items that still have value, these firms will buy the excess IT hardware and sell refurbished products to those seeking more affordable options than buying new.

Corporations face risks both for incompletely destroyed data and for improperly disposed computers and, according to the RCRA, are liable for compliance with regulations even if the recycling process is outsourced. Companies can mitigate these risks by requiring waivers of liability, audit trails, certificates of data destruction, signed confidentiality agreements, and random audits of information security. The National Association for Information Destruction (NAID) is an international trade association for data destruction providers.

Secure recycling   There are regulations that monitor the data security on end-of-life hardware. NAID is the international trade association for companies providing information destruction services. Suppliers of products, equipment, and services to destruction companies are also eligible for membership. NAID's mission is to promote the information destruction industry and the standards and ethics of its member companies.

The typical process for computer recycling aims to securely destroy hard drives while still recycling the by-product. A typical process for effective computer recycling accomplishes the following:

1 Receive hardware for destruction in locked and securely transported vehicles.
2 Shred hard drives.
3 Separate all aluminum from the waste metals with an electromagnet.
4 Collect and securely deliver the shredded remains to an aluminum recycler.
5 Mold the remaining hard drive parts into aluminum ingots.

Compressed Air

Compressed air is used in almost every sector of the economy. It is regarded as the fourth utility, after electricity, natural gas, and water. In comparing it to electric motors, compressed air produces smooth translation with much more uniform force. Compressed air equipment can be more economical and more durable. Perhaps the greatest advantage of compressed air, however, is the high ratio of power to weight or power to volume.

However, per unit of energy delivered, compressed air is more expensive than the other three utilities. In Europe 10 percent of all electricity used by industry is used to produce compressed air. This amounts to 80 terawatt hours per year. In most plants, it also represents a substantial opportunity for reducing plant energy consumption and cutting operating costs. Consider that compressed air:

- Is the most inefficient source of energy in a plant
- Is often the biggest end use of electricity in a plant
- Is often used inappropriately
- Is a system that can be better managed
- Has costs that can be measured
- May have an overall system efficiency as low as 15 percent

COMPRESSOR SELECTION

A general rule of thumb is that centrifugal and rotary compressors are better suited for continuous base load type of service. Reciprocating air compressors are better suited for intermittent applications or swings in load. Incorrect selection of compressor controls and incorrect operation can lead to excessive energy consumption.

Compressed air is air that is kept under a certain pressure, usually greater than that of the atmosphere.[7]

AUDIT CHECKLIST

Potentially inappropriate applications   Compressed air should not be used for the following inappropriate and wasteful applications:

■ Open blowing
■ Sparging (agitating, stirring, mixing)
■ Aspirating
■ Atomizing
■ Padding
■ Dilute phase transport
■ Dense phase transport
■ Vacuum generation
■ Personnel cooling
■ Open handheld blowguns or lances
■ Cabinet cooling
■ Vacuum venturis
■ Diaphragm pumps
■ Timer drains or open drains
■ Air motors

Compressed air system growth in a plant is often the result of incremental plant growth (expansion of a plant and/or additions of equipment within a plant over time). The nature of this incremental growth is that the system in a plant today may not resemble what was intended 25 years ago when a plant was initially built. There are often opportunities to optimize a system's operation today through analysis of the supply and demand sides.[8]

System assessment
■ Identify lowest optimum target pressure
■ Resolve pressure profile and control issues
■ Validate perceived high-pressure uses
■ Determine air storage for high volume intermittent demand
■ Resolve piping deficiencies and eliminate gradients

EFFICIENCY MEASURES

Air leak reduction tips
■ Improve end use efficiency
■ Reduce system air pressure
■ Use unloading controls
■ Adjust cascading set points
■ Use automatic sequencer

- Reduce run time
- Add primary receiver volume

System pressure tips
- The system should be delivering at the lowest practical pressure.
- A properly designed system should have a pressure loss of less than 10 percent of the compressor discharge pressure.
- A 2-pound per square inch (1406-kilograms per square meter) increase in system pressure can increase energy costs by 1 percent.
- The leading cause of pressure drop is a poorly designed distribution system.

Maintenance
Inadequate maintenance can have a significant impact on energy consumption via lower compression efficiency, air leakage, or pressure variability. In many cases, it makes more sense from efficiency and economic standpoints to maintain equipment more frequently than the intervals recommended by manufacturers, which are primarily designed to protect equipment.

End use filters, regulators, and lubricators should be checked during a plant's periodic preventative maintenance review. Filters should be inspected more often because a clogged filter will increase pressure drop, which either can reduce pressure at the point of use or increase the pressure required from the compressor, thereby consuming excessive energy.[9]

Endnotes

1. Hubbard, Roy S. "Understanding Water-Chiller Efficiency Ratings Evaluating Capital Cost and Energy Efficiency." http://www.fmlink.com/ProfResources/Sustainability/Articles/article.cgi?USGBC:200712-21.html. Accessed April 12, 2010.
2. Brooks, Nate and Adan Reinosa."Efficiency Estimation and Payback Periods - the Impact for Industrial Electric Motor Users." http://www.plantservices.com/wp_downloads/080801_Baker_Inst_Co.html. Accessed March 10, 2010.
3. Puskar, John."20 Tips for Reducing Fuel Costs in Ovens and Furnaces." http://www.fmlink.com/ProfResources/Magazines/article.cgi?AFE:afe041508a.html. Accessed March 10, 2010.
4. E24™ Powered Roller Conveyors. http://www.cisco-eagle.com/systems/conveyors/conveyor/e24-power-roller-conveyor.htm. Accessed September 10, 2010.
5. Bott, Dan. "Improve energy efficiency by restructuring your vacuum generators." http://www.plantservices.com/articles/2005/530.html?page=print. Accessed February 12, 2010.
6. Roseman, Bob. "There is more than one way to dry the air." http://www.plantservices.com/articles/2006/202.html. Accessed March 10, 2010.

7. Scott, Jeff. Trident Compressed Air. "Industrial Compressed Air Systems." http://www
.chatham-kent.ca/cityBundle_services/downloadsService/downloadfiles/58b3462b
-a40b-4ab8-9429-062a177817b5_Trident%20Compressed%20Air.pdf. Accessed April
12, 2010.
8. Smith, Rod. "Sustainability at RR Donnelley." http://www.airbestpractices.com/
energy-manager/sustainable-manufacturing-news/sustainability-rr-donnelley.
Accessed March 10, 2010.
9. U.S. Department of Energy. "Compressed Air Best Practices Tools." http://www1.eere
.energy.gov/industry/bestpractices/pdfs/compressed_air_webcast_0307.pdf.
Accessed April 12, 2010.

7

EXTERIOR

Landscaping

Sustainable landscaping encompasses a variety of practices that have developed in response to environmental issues. These practices are used in every phase of landscaping, including design, construction, implementation, and management of residential and commercial landscapes.

SUSTAINABLE LANDSCAPING PRACTICES

Some sustainable landscape solutions are as follows:

- Reduction of stormwater runoff through the use of bio-swales, rain gardens, and green roofs and walls
- Reduction of water use in landscapes through design of water-wise garden techniques (sometimes known as xeriscaping)
- Bio-filtering of wastes through constructed wetlands
- Landscape irrigation using water from showers and sinks, known as gray water
- Integrated pest management techniques for pest control
- Creating and enhancing wildlife habitat in urban environments
- Energy-efficient landscape design in the form of proper placement and selection of shade trees and creation of wind breaks
- Permeable paving materials to reduce stormwater runoff and allow rain water to infiltrate into the ground and replenish groundwater rather than run into surface water
- Use of sustainably harvested wood, composite wood products for decking and other landscape projects, as well as use of plastic lumber
- Recycling of products, such as glass, rubber from tires, and other materials, to create landscape products such as paving stones, mulch, and other materials
- Soil management techniques, including composting kitchen and yard wastes, to maintain and enhance healthy soil that supports a diversity of soil life

Tip Box

TABLE 7.1   SUSTAINABLE SITE PRACTICES
■ Xeriscaping or drought-resistant landscaping
■ Permeable paving
■ Shielded light fixtures
■ Stormwater bio-filtering
■ Stormwater breaks to reduce flow
■ Gray water landscape irrigation
■ Creation of site rain gardens
■ Landscape bed mulching
■ Landscaping used as windbreaks
■ Compost additives to soil
■ Landscaping used as building shading
■ Green roofs used to reduce runoff

■ Integration and adoption of renewable energy, including solar-powered landscape lighting

See Table 7.1 for a summary of site sustainability practices.

SHADING AND WINDBREAKS

Plants used as windbreaks can save up to 30 percent on heating costs in winter. They also help with shading a commercial building in the summer, creating cool air around the perimeter of the building through evapotranspiration. Shading can also be beneficial in cooling parking lots, sidewalks, and other hardscape site areas. See Table 7.2 for a list of common United States trees useful for shading buildings and parking lots (note: consult a local arborist for appropriate trees in a region).

COMPOST ADDITIVES

Compost can be added as an amendment to poorly draining soil, as a fertilizer on flower and vegetable beds, and to fruit trees or used as a potting soil for potted plants. Trimmings from lawns, trees, and shrubs from a large landscape site can be used as feedstock for on-site composting. Reusing on-site organic materials will decrease the need for purchasing other soil additives.

MULCH AND WATER REDUCTION

Using mulch is an effective way to reduce water loss due to evaporation, reduce weed growth, and minimize erosion, dust, and mud problems. Mulch will also add nutrients to the soil as it decomposes. Mulch beds hold water better than unmulched beds.

TABLE 7.2   COMMON URBAN TREES FOR SHADING BUILDING WALLS AND PARKING AREAS

FAMILY	COMMON NAME	CALIPER	HEIGHT	CANOPY
Celtis australis	European hackberry	6"–8"	35'–40'	large
Celtis occidentalis	Common hackberry	6"–8"	35'–40'	large
Celtis sinensis	Chinese hackberry	6"–8"	35'–40'	large
Cinnamomum camphora	Camphor	8" +	35'–40'	large
Fraxinus americana	Autumn purple ash	6"–8"	35'–40'	large
Fraxinus americana	Rosehill ash	6"–8"	35'–40'	large
Fraxinus excelsior	Hessei ash	6"–8"	35'–40'	large
Ginkgo biloba	Ginkgo	4"–6"	25'–30'	large
Princeton Sentry	Princeton sentry	6"–8"	50'–70'	medium
Liriodendron tulipifera	Tulip tree	6"–8"	35'–40'	large
Magnolia grandiflora	Southern magnolia	4"–6"	25'–30'	large
Pistacia chinesis	Chinese pistache	4"–4"	25'–30'	medium
Pyrus kawakamii	Evergreen pear	4"–6"	25'–30'	medium
Quercus agrifolia	Coastal live oak	6"–8"	35'–40'	large
Quercus bicolor	Swamp white oak	6"–8"	35'–40'	large
Quercus calliprinos	Palestine live oak	6"–8"	35'–40'	large
Quercus castaneafolia	Chestnut-leaved oak	6"–8"	35'–40'	large
Quercus chrysolepis	Canyon live oak	6"–8"	35'–40'	large
Quercus frainetto	Forest green oak	6"–8"	35'–40'	large
Quercus lobata	Valley oak	6"–8"	35'–40'	large
Quercus muehlenbergii	Chinkapin oak	6"–8"	35'–40'	large
Quercus nigra	Water oak	6"–8"	35'–40'	large
Quercus phellos	Willow oak	6"–8"	35'–40'	large
Quercus rubra	Red oak	6"–8"	35'–40'	large
Quercus shumardii	Shumard oak	6"–8"	35'–40'	large
Quercus suber	Cork oak	6"–8"	35'–40'	large
Quercus virgianiana	Southern live oak	6"–8"	35'–40'	large
Quercus wislizenii	Interior live oak	6"–8"	35'–40'	large
Zelkova serrata	Green vase	6"–8"	35'–40'	large
Zelkova serrata	Musashino	6"–8"	35'–40'	large
Zelkova serrata	Village green	6"–8"	35'–40'	large

XERISCAPING

A native plant, that has adapted to local climate conditions, will require less work and cost to maintain. Selecting the right kind of local plants can minimize water consumption and maintenance needs. Companies should consult with a landscape architect to either install a drought-resistant landscaping package that requires little water, or use a native plants design that is adapted to normal rainfall totals for the area.

LANDSCAPE MAINTENANCE

General annual landscape maintenance tasks:
- Annually aerate lawn areas
- Reapply mulch as necessary
- Fertilize as directed
- Remove dead plant debris
- Prune woody plants

Detailed landscape design and maintenance strategies:

Water conservation
- Use trees and shrubs that are native to the area or are drought-resistant.
- Plant drought-resistant turfs and grasses that require less water.
- Reduce water consumption by using low-flow, micro-spray, or drip irrigation systems that reduce water use and loss through evaporation.
- Use automatic rain shutoff sprinkler systems and intelligent irrigation systems that use soil moisture sensors.
- Arrange or adjust sprinkler heads to prevent overspray onto sidewalks or streets.
- Place 2–3 inches (5.1–7.6 centimeters) of environmentally friendly mulch around shrubs, trees, and plant beds to retain moisture.
- Install separate irrigation zones for lawn and landscape plants, or utilize drought-resistant plants that do not require irrigation.
- Group plants with similar water requirements in the same area.
- Avoid plants that are not well adapted to the local climate.
- Use rain barrels, downspouts, and gutters to direct and collect water that can be used for the irrigation of plants.
- Take care with the landscape design to avoid placing trees and shrubs in environmental conditions where the lighting, emissions, moisture, or heat may cause the plant to become overstressed or prone to problems.
- Create swales and rain gardens to filter site runoff. Use berms and swales to provide proper site drainage and grade control.
- Create slopes that can be terraced with a series of raised beds or planters. Avoid steep banks or retaining walls.
- Adhere to new fertilizer and herbicide ordinances by the municipality, state, or U.S. Environmental Protection Agency.

Water quality

■ Reduce soil additives and fertilize only when necessary.

■ Restrict fertilizer use on landscaped areas near shorelines, bays, estuaries, and waterways.

■ Calculate fertilizer applications based on soil analysis and specific plant fertilizer needs, rather than on standard recommended application rates.

■ Use fertilizers that are slow or time-released nitrogen, with reduced phosphorus content, to prevent runoff of nutrients into waterways.

■ Reduce stormwater runoff and absorption rates by using products such as crushed shell, gravel, and mulch for sidewalks, driveways, and footpaths.

■ Plant eco-friendly, drought-resistant plants that can absorb harmful runoff from nutrients (nitrogen and phosphorus).

■ Use hardscape or paver materials that are environmentally friendly and permeable. An example is permeable interlocking concrete pavers laid atop a layer of crushed stone that allows water to infiltrate back into the soil subgrade. This type of installation allows stormwater to be filtered and reduces pollutants.

Resource conservation

■ Utilize mulches that are recycled or by-products of other materials that are environmentally sustainable, such as pine needles, tree bark, and leaves.

■ Place organic yard wastes in compost areas scattered across the site. Yard wastes such as grass clippings, leaves, vegetable and flower plants, and small amounts of woody material can be used as soil amendments and to reduce the amount of material going to landfills.

■ Remove invasive plant species that crowd out native plants.

■ Select trees that are beneficial for wind protection. Do not plant trees too close to homes or structures to protect them from fires or lightening strikes.

■ Select plants that do not require very much pruning or create yard waste.

■ Reuse materials such as broken concrete or recycled plastic material to erect retaining walls to create visual features instead of removing it to the landfill.

■ Avoid using building salvage or other materials that may be harmful to the soil or dangerous to food or crops, such as creosote-stained or chemically treated railroad ties to build raised beds.

Energy conservation

■ Select trees that can offer shading to help cool buildings, particularly on the south side to reduce energy consumption. The southern exposures of buildings receive the most intense sunlight, while east and west exposures receive morning and afternoon sunlight. The north side receives the least sunlight.

■ Use shade trees for parking areas or other hard surface areas.

■ Shade air conditioning and cooling units so they will run cooler with increased energy efficiency.

■ Use solar power for site lighting, where possible, to reduce energy consumption. Use low solar landscape lighting versus electrical high- or low-voltage lighting. Solar lights are typically dimmer than wired landscape lighting but do not use consumable energy.

■ Incorporate site lighting that reduces light pollution, reduces excessive glare, prevents light trespassing on others, improves security and safety, and reduces energy consumption.

Diversity and wildlife habitats

■ Many landscape pest problems can be traced directly to plant health problems. Including a diverse range of plants on the site will help maintain beneficial organism populations.

■ Examples of helpful organisms are birds, reptiles, small animals, insects, and microorganisms. Since most plant pests target a certain species or family of plants, a diversity of plants is the best way to control pest problems. Predators and parasites of plant pests are also considered beneficial organisms.

■ Incorporate in the landscape design bird houses, fountains, and ponds as drinking areas for wildlife.

■ Incorporate in the landscape design trees, shrubs, and plants that offer nesting areas and food for birds, and possible habitats for other small wildlife.

■ Use plants, where appropriate, that attract the honeybees, which have been compromised recently. Plants that promote bee colony maintenance include citrus trees, magnolia, holly, shrubs like privets and ligustrums, and glossy abelia.

■ Where appropriate, consider incorporating a community garden into the site. Community gardens (or simply a collection of fruit-bearing trees) include sustainable and edible landscape opportunities such as fruit trees, vegetables, and herbs that reduce food costs, add to the safety of where foods are grown, and are convenient and fun to grow.

Maintenance

■ Manage weed and pest problems with environmentally friendly, organic solutions.

■ Utilize landscape contractors and lawn maintenance companies knowledgeable about sustainable landscape methods and who have been certified according to Best Management Practices by the state, municipality, or independent organizations.

■ Annually aerate lawn areas, reapply mulch as necessary, fertilize as required, remove dead plant debris, and prune woody plants.[1]

Paving and Surfacing

Sustainable paving, or permeable paving, helps mitigate the negative impacts of conventional paving by reducing stormwater runoff, improving water quality, reducing flooding, and recharging groundwater. It can be part of an overall strategy to improve the environmental performance of developed areas. Using permeable paving in parking and pedestrian areas helps reduce the temperature, volume, and velocity of stormwater runoff. It also filters and removes pollutants, and creates more sustainable communities.

Permeable paving is comprised of a range of sustainable materials and techniques used on roads, bike paths, parking lots, and pavements that allow the movement of water and air around the paving material. Although some porous paving materials look virtually identical to nonporous materials, their environmental effects are quite different. Pervious products (whether concrete, porous asphalt, stones, or bricks) allow precipitation to percolate through the material and into an underlying substrate (usually an underground stone storage bed) designed to accommodate the water. The result is that paved or hard surface areas that would normally generate stormwater runoff, and require site accommodations to accommodate the stormwater, will generate little or no runoff.

TYPES OF PERMEABLE PAVING

Installation of porous pavements is no more difficult than that of dense pavements, but has different specifications and procedures that must be strictly adhered to. Eight general types of porous paving materials present distinctive advantages and disadvantages for specific applications. Here are examples of permeable paving products:

1 **Pervious concrete:** Pervious concrete is widely available, can bear frequent traffic, and is universally accessible. Pervious concrete quality depends on the installer's knowledge and experience.

2 **Porous asphalt:** Porous asphalt is mixed at conventional asphalt plants, but fine (small) aggregate is omitted from the mixture. The remaining large, single-sized aggregate particles leave open voids that give the material its porosity and permeability. Under the porous asphalt surface is a base course of additional single-sized aggregate. Porous asphalt surfaces are being used on highways to improve driving safety by removing water from the surface.

3 **Single-sized aggregate:** Single-sized aggregate without any binder, e.g., loose gravel, stone-chippings, is the most permeable paving material in existence, and the least expensive. Although it can only be safely used in very low-speed, low-traffic settings, e.g., car-parks and drives, its potential cumulative area is great.

4 **Porous turf:** Porous turf, if properly constructed, can be used for occasional parking like that at churches and stadiums. Plastic turf reinforcing grids can be used to support the increased load. Living turf transpires water, actively counteracting the "heat island" with what appears to be a green open lawn.

5 **Permeable interlocking pavers:** Permeable interlocking pavers, available in a variety of colors, styles, and patterns, are an attractive alternative to asphalt or concrete and provide greater design flexibility. These systems consist of interlocking concrete pavers with a permeable material in the voids, or joints, between the pavers. Water seeps through these joints and infiltrates into a stone base and eventually into the ground. Interlocking paving systems effectively provide infiltration, detention, and treatment of stormwater runoff. They are applicable to small or large paved areas and offer flexibility in surface repair since sections can be removed and replaced. Recent advances allow installation of larger modules—50 to 100 individual pavers to reduce labor installation costs. Pavers provide an architectural appearance, and can bear surprisingly heavy traffic. Interlocking concrete pavers are appropriate for

light traffic (though not high-volume or high-speed roads), and include a variety of porous products that reduce site rainwater runoff.

6 **Resin bound paving:** Resin bound paving is a mixture of resin binder and aggregate. Clear resin is used to fully coat each aggregate particle before laying. Enough resin is used to allow each aggregate particle to adhere to one another and to the base yet leave voids for water to permeate through. Resin bound paving provides a strong and durable surface that is suitable for pedestrian and vehicular traffic in applications such as pathways, driveways, car-parks, and access roads.

7 **Bound recycled glass porous pavement:** Bound recycled glass porous pavement consists of bonding processed postconsumer glass with a mixture of resins, pigments, and binding agents. The product provides a permeable paving material that also reuses materials that would otherwise be disposed in landfills. Approximately 75 percent of glass in the United States is thrown away in landfills.

8 **Reinforced grass/gravel paving:** Reinforced grass/gravel paving is appropriate for low-use facilities, such as emergency lanes, occasional-use parking, weekend markets, and events. This system consists of a stable base (grid framework), usually constructed of concrete or plastic, that contains voids for the placement of turf grass or gravel. A stone or sand drainage system is placed under the framework for stormwater management. This type of paving supports frequent light and occasional heavy loads without damaging the structure and the underlying soil. Grass and gravel dissipate heat and are an effective alternative to conventional paving in appropriate conditions. This system has been used in underutilized parking areas for schools, universities, and stadiums and in alleys and driveways to create more hospitable environments.

Of the paving materials listed above, asphalt and concrete are by far the most common paving materials used in commercial facilities. Porous asphalt or concrete is made without fine particles and differs in this way from conventional paving. Without these fine particles, the paving material contains more voids, or air spaces, allowing water to pass through the pavement into a reservoir of crushed stone and then into the ground. This results in a mix having a consistency of popcorn or rice cakes. Porous asphalt and concrete can be mixed and installed on-site in the same manner as conventional paving, although installation requires more expertise. When the finer particles are removed, these surfaces lose some of their strength. For this reason, permeable paving is not currently recommended for high-traffic areas, such as roadways, or high- or heavy-traffic parking lots. It is typically recommended for low-use areas where load-bearing or weight issues are less critical, such as pedestrian walkways, parking bays, secondary or light-duty parking areas, or in low-traffic areas, such as driveways and loading areas.[2]

Stormwater

BENEFITS OF NATURAL STORMWATER MANAGEMENT

Controlling and reducing the amount of runoff from a site can, in turn, reduce the risk of flooding and the incidence of combined sewer overflow (CSO) events, as well as reduce the size and extent of drainage infrastructure. Retention and detention systems may also decrease overall water consumption as collected rainwater and snowmelt can be reused on-site.

VEGETATION

The use of vegetation in sustainable stormwater management practices provides substantial benefits for water quality and quantity. Vegetation intercepts rainfall, reducing the erosivity of raindrops. Vegetation increases surface friction, slowing runoff and promoting settling of suspended solids. Root systems increase pore space and promote infiltration of stormwater, and the roots themselves help to anchor soil, protecting against rill and gully formation from concentrated flows. Finally, vegetation can help to remove pollutants from stormwater through biological uptake of nutrients.

Sustainable stormwater best management practices can also offer cost savings over traditional stormwater management approaches. On-site infiltration or detention of rainwater can have a direct effect on the size and number of drainage pipes needed for handling runoff, resulting in lower construction costs.

Green spaces incorporated into a company's site design and stormwater management plan are highly valued and can increase property values and enhance tourism opportunities. Some highly visible and publicized installations result in national attention, promoting the city or organization responsible for their development. Businesses can demonstrate their commitment to the environment by incorporating sustainable features into their corporate campuses and other locations.

Sustainable practices on individual sites can carry area-wide benefits. Reduction in the volume and velocity of stormwater entering local water bodies reduces stream erosion on- and off-site. Preservation of open spaces allows for habitat preservation and restoration, and installations can be designed to support a broad range of landscape types. The focus on reconnecting the natural hydrologic cycle can preserve the integrity of natural systems, minimizing disturbance and potentially returning areas to predevelopment conditions.

STORMWATER MANAGEMENT STRATEGIES

See Table 7.3 for a summary of site stormwater management strategies.

- **Downspout disconnection:** In urban areas, downspouts are commonly connected to drain tiles that feed the sewer system, and the cumulative effect of thousands of connected downspouts can greatly increase the annual number, magnitude, and duration of CSO events. Downspout disconnection is the process of separating roof downspouts from the sewer system and redirecting roof runoff onto pervious surfaces, most commonly a lawn. This reduces the amount of directly connected impervious area in a drainage area.
- **Filter strips:** Filter strips are bands of dense, permanent vegetation with a uniform slope, primarily designed to improve water quality between a runoff source, such as a parking lot or roof, and the area where the water is collected or absorbed. Filter strips can substantially decrease the amount and rate of runoff exiting a site.

TABLE 7.3 STORMWATER MANAGEMENT STRATEGIES

- Disconnect downspouts to reduce flow rate
- Install filter strips for pretreatment of runoff
- Use dry wells and infiltration trenches to percolate runoff
- Develop pocket wetlands to control volume and treat pollutants
- Use porous paving to reduce runoff quantity
- Install rain barrels and cisterns to store runoff for site use
- Develop rain gardens to store and infiltrate stormwater
- Use soil amendments to increase water retention of soil
- Install tree box filters to contain water for use by trees
- Install green roofs to reduce roof runoff quantity
- Use vegetated swales to control and infiltrate runoff

- **Infiltration practices:** Infiltration practices are designs that enhance water percolation through a media matrix that slows and partially holds stormwater runoff and facilitates pollutant removal. Examples include dry wells and infiltration trenches.
- **Pocket wetlands:** Pocket wetlands are constructed shallow marsh systems designed and placed to control stormwater volume and facilitate pollutant removal. As engineered constructed facilities, pocket wetlands have less biodiversity than natural wetlands but still require a base flow to support the aquatic vegetation present. Pollutant removal in these systems occurs through the settling of larger solids and coarse organic material and also by uptake in the aquatic vegetation. Pocket wetlands are designed with three distinct zones: a break immediately after the inlet to receive stormwater, the wetland area, and a micropool immediately prior to the outfall. The forebay and micropool allow for sediment control.
- **Porous pavement:** See the pavement section of this chapter for information regarding this stormwater runoff strategy.
- **Rain barrels/cisterns:** Rain barrels and cisterns collect building runoff from roof downspouts and store it for later reuse for nonpotable applications, such as irrigation. Rain barrels are most often used for individual residences, while cisterns have both residential and commercial applications. Both storage devices act to decrease the volume and flow rate of rooftop-generated stormwater runoff. Rain barrels and cisterns can provide a source of chemically untreated "soft water" for gardens and compost and other nonpotable needs, free of most sediment and dissolved salts.
- **Rain gardens:** Rain gardens, also known as bioretention cells, are vegetated depressions that store and infiltrate runoff. Uptake into plants reduces runoff volume and pollutant concentrations. The soil media is engineered to maximize infiltration and pollutant removal. Rain gardens are typically designed to avoid ponding for longer than 24 hours.

- **Soil amendments:** Including the use of both soil conditioners and fertilizers, soil amendments make the soil more suitable for the growth of plants and increase water retention capabilities. The use of soil amendments is conditional on their compatibility with existing vegetation. A variety of techniques are included as potential soil amendments including aerating, fertilizing, and adding compost, other organic matter, or lime to the soil. See the soil amendments section of this chapter for additional information.

- **Tree box filters:** Tree box filters are inground containers typically containing street trees in urban areas. Runoff is directed to the tree box, where it is filtered by vegetation and soil before entering a catch basin. Tree box filters adapt bioretention principles used in rain gardens to enhance pollutant removal, improve reliability, standardize and increase ease of construction, and reduce maintenance costs.

- **Green roofs (vegetated roofs or roof gardens):** Green roofs, also known as eco-roofs or nature roofs, are structural components that help to mitigate the effects of urbanization on water quality by filtering, absorbing, or detaining rainfall. Modern vegetated roofs can be categorized as "intensive" or "extensive" systems depending on the plant material and planned usage for the roof area. Intensive vegetated roofs utilize a wide variety of plant species that may include trees and shrubs, require deeper substrate layers [usually 4 inches (10 centimeters)], are generally limited to flat roofs, require "intense" maintenance, and are often park-like areas accessible to the general public. In contrast, extensive roofs are limited to herbs, grasses, mosses, and drought-tolerant succulents, such as *Sedum*, can be sustained in a shallow substrate layer less than 4 inches (10 centimeters), require minimal maintenance, and are generally not accessible to the public.

- **Vegetated swales:** Vegetated swales are broad, shallow channels designed to convey and infiltrate stormwater runoff. The swales are vegetated along the bottom and sides of the channel, with side vegetation at a height greater than the maximum design stormwater volume. The design of swales seeks to reduce stormwater volume through infiltration, improve water quality through infiltration and vegetative filtering, and reduce runoff velocity by increasing flow path lengths and channel roughness. Two types of swales exist. Dry swales are designed with highly permeable soils and an underdrain to allow the entire stormwater volume to convey or infiltrate away from the surface of the swale shortly after storm events. Dry swales may be designed with check dams that act as flow spreaders and encourage sheet flow along the swale. Check dams also retain stormwater. Wet swales are designed to retain water and maintain marshy conditions for the support of aquatic vegetation. Because of their highly permeable soil and conveyance capability, dry swales are more applicable for urban environments.[3]

Shading

SHADING BENEFITS

During cooling seasons, external window shading is an excellent way to prevent unwanted solar heat gain from entering a conditioned space. Shading can be provided by natural landscaping or by building elements, such as awnings, overhangs, and trellises. Some shading devices can also function as reflectors, called light shelves, which bounce natural light for daylighting deep into building interiors. See Fig. 7.1 for a photograph of horizontal shading elements on an office building.

The design of effective shading devices relies on the solar orientation of a particular building facade. For example, simple fixed overhangs are very effective at shading south-facing windows in the summer when sun angles are high. However, the same horizontal device is ineffective at preventing low afternoon sun from entering west-facing windows during peak heat gain periods in the summer.

Exterior shading devices are particularly effective in conjunction with clear glass facades. However, high-performance glazings are now available that have very low shading coefficients (SC). High-performance glazing products reduce the need for exterior shading devices.

Solar control and shading can be provided by a wide range of building components including:

- Landscape features, such as mature trees or hedge rows
- Exterior elements, such as overhangs or vertical fins

Figure 7.1   Horizontal shading elements on an office building.

- Horizontal reflecting surfaces called light shelves
- Low SC glass
- Interior glare control devices, such as venetian blinds or adjustable louvers

Fixed exterior shading devices such as overhangs are generally most practical for small commercial buildings. The optimal length of an overhang depends on the size of the window and the relative importance of heating and cooling in the building.[4]

GENERAL SHADING RECOMMENDATIONS

Architectural

1 Use fixed overhangs on south-facing glass to control direct beam solar radiation. Indirect (diffuse) radiation should be controlled by other measures, such as low-emissivity glazing.
2 To the greatest extent possible, limit the amount of east and west glass since it is harder to shade than south glass. Consider the use of landscaping to shade east and west exposures.
3 Do not worry about shading north-facing glass in the continental United States latitudes since it receives very little direct solar gain. In the tropics, disregard this rule of thumb since the north side of a building will receive more direct solar gain. Also, in the tropics consider shading the roof even if there are no skylights since the roof is a major source of transmitted solar gain into the building.
4 Since shading affects daylighting, designers must consider both at the same time. For example, a light shelf bounces natural light deeply into a room through high windows while shading lower windows.
5 Do not expect interior shading devices such as venetian blinds or vertical louvers to reduce cooling loads since the solar gain has already been admitted into the work space. However, these interior devices do offer glare control and can contribute to visual acuity and visual interest in the work place.
6 Pay attention to area sun angles. An understanding of sun angles on the site is critical to creating an energy-efficient design. Aspects of sun angle include determining basic building orientation, selecting shading devices, and the placement of photovoltaic panels or other types of solar collectors.
7 Durability is a major factor in the selection of shading devices for a building. Over time, operable shading devices can require a considerable amount of maintenance and repair.
8 When relying on landscape elements for shading, be sure to consider the cost of landscape maintenance and upkeep on life-cycle cost.
9 Shading strategies that work well at one latitude may be completely inappropriate for other sites at different latitudes. Be careful when applying shading ideas from one project to another.

HVAC Good shading provides cooling load reductions that can be calculated and incorporated into the HVAC system sizing. The mechanical engineer should perform calculations that include shaded windows, but acknowledge that not all shading systems will be deployed when needed.

Daylighting   Shading devices modify the intensity and distribution of daylight entering the space. Lighting design scheme and placement of control zones may be affected. Shading devices can be designed as part reflectors, blocking direct sunlight from entering the space and affecting the heating load, but utilizing reflected light to provide daylighting benefits.

Cost-effectiveness   Proper shading devices can be partially or fully paid for by reduced HVAC equipment costs. However, the likelihood of proper use by occupants must be accounted for. Building managers should estimate the savings, and calculate a return on investment for the cost of installing shading devices.

EXTERIOR SHADING STRATEGIES

- Use exterior shading, either a device attached to the building skin or an extension of the skin itself, to keep out unwanted solar heat. Exterior systems are typically more effective than interior systems in blocking solar heat gain.
- Design the building in such a way that it provides shade for itself. If shading attachments are not aesthetically acceptable, designers can use the building form itself for exterior shading. Examples include setting the window back in a deeper wall section, or extending elements of the building skin to visually blend in with envelope structural features.
- Where possible, use horizontal shading devices to shade south windows. Examples include awnings, overhangs, screen walls, horizontal projections, and recessed windows.
- Shading is somewhat useful on the east and west sides of a building. Horizontal shading devices serve no useful function on the north side of a building.
- Use vertical shading on east and west windows. Examples include vertical fins or recessed windows.
- Vertical fins may be somewhat useful on the north side of a building to block early morning and late afternoon low sun.
- Give shading priority to the west and south windows of a building. Morning sun is usually not a serious heat gain problem. If the retrofit or construction budget is tight, invest in west and south shading to achieve the greatest effect for the money spent.

See Table 7.4 for a summary of exterior building shading strategies.

WINDOW UNIT SHADING

- Use exterior window shades for a smooth facade. Exterior shade screens are highly effective on all facades and permit filtered view, though maintenance and durability factors must be considered as well.

Tip Box

TABLE 7.4   EXTERIOR SHADING STRATEGIES

- Shading devices can be the building skin or an attachment.
- Adjacent large trees can be effective shading tools.
- Shading attachment types:
  - Awnings
  - Shelves or overhangs
  - Green wall systems
- Recessed windows can be incorporated into designs to provide shading.
- Building offsets may be used to shade portions of the building.
- Use horizontal shading on the south side of the building.
- Use vertical shading on the east and west sides.
- Shading is not necessary on the north side.
- South and west shading should take priority.

- Manufacturers make exterior roller shades for a movable alternative. Opaque shades, of course, are very effective while open weave shades are not. Exterior roller shades may be appropriate for low facilities with few windows, but would represent an aesthetic and maintenance headache for larger buildings with a greater amount of glazing.
- Avoid relying on dark glazing. Glazing treatments (reflective coatings, heavy tints, and reflective retrofit film) can be effective at reducing heat transfer, since they allow direct sun penetration but with reduced intensity. This may not be an effective shading strategy from an occupant's perspective unless the transmittance is very low to control glare, for example, 5 to 10 percent. Fritted glass, with a durable diffuse or patterned layer fused to the glass surface, can also provide some degree of sun control, depending upon the coating and glass substrate properties. This type of glazing may also increase glare, however.
- Consider between-glass shading systems. Several manufacturers offer shading systems (for example, blinds) located between layers of glazing. Some of these systems are fixed while others are adjustable.[5]

Cladding and Siding Materials

Exterior cladding is generally defined as a protective layer or finish affixed to the exterior side of a building enclosure system. The word *cladding* is often used as a general reference to a wide variety of naturally occurring and synthetic, or man-made, building envelope materials, components, and systems. Typically, these elements are quarried, manufactured, or otherwise developed and/or altered to render them suitable for use on the exterior of a building or structure. These are frequently derived from, or tailored to, the available resources, raw materials, and climatic conditions unique to a particular geographic region or exposure. Exterior cladding is generally the first, though not necessarily the primary, line of defense against bulk rainwater penetration.

BASIC EXTERIOR WALL TYPES

Wall cladding products and systems come in a dizzying array of styles, colors, and configurations. Manufacturers routinely claim sustainability benefits for their products, and tie their attributes to Leadership in Energy and Environmental Design (LEED) credit contributions or energy-saving attributes. As a component in an overall system, most building cladding products can truthfully make sustainability claims of some sort. Designers and building managers must examine the choice of exterior wall cladding in the context of the environment they build in, the building occupancy, and the energy characteristics of the overall wall construction (including insulation, mass, and structure).[6]

Wall Cladding and Wall Systems (generic types)
- Cast-in-place concrete
- Exterior Insulation and Finish System (EIFS)
- Masonry (brick and concrete masonry units)
- Panelized metal
- Precast and prestressed concrete
- Thin stone (natural stone and synthetic stone)
- Wood
- Fiber cement
- Vinyl

Exterior wall types commonly associated with above-grade, commercial building enclosure design and construction in the United States can generally be classified as follows: a cavity wall, a barrier wall, or a mass wall. Following is a summary of the characteristics of each wall type:

Cavity wall   A cavity wall (also referred to as "screen" or "drained" wall systems) is considered by many to be the preferred method of construction in most climatic and rainfall zones in the United States. This term is now used more generically to define any wall system or assembly that relies upon a partially or fully concealed air space and drainage plane to resist bulk rainwater penetration and, depending upon the design, to improve the overall thermal performance at the building enclosure.

Barrier wall   As the name implies, this term is commonly used to describe any exterior wall system of assembly that relies principally upon the weather-tight integrity of the outermost exterior wall surfaces and construction joints to resist bulk rainwater penetration and/or moisture ingress. This type of wall system is commonly associated with precast concrete spandrel panels, certain types of composite and solid metal plate exterior cladding systems, and early-generation EIFS.

Mass wall   Unlike a cavity wall system, where the wall is constructed with a wall cavity and through-wall flashing to collect and redirect bulk rainwater to the building exterior, mass walls rely principally upon a combination of wall thickness, storage capacity, and (in masonry construction) bond intimacy between masonry units and mortar to effectively resist bulk rainwater penetration. For economic reasons, mass walls are less common in design and construction today.

COMMON ELEMENTS OF AN EXTERIOR WALL

Each of the above wall types, or combination thereof, generally consists of the following basic elements, or layers:

- Exterior cladding (natural or synthetic)
- Drainage plane
- Air barrier system
- Vapor retarder
- Insulating element
- Structural element

Several of these layers may, at the discretion of the design professional, serve multiple purposes. For example, in barrier wall design and construction, the exterior cladding material may be designed to function both as the primary drainage plane and principal air barrier for a building or structure. Similarly, in cavity wall construction, rigid insulation placed inside the exterior wall cavity, if properly detailed, may also function as the air barrier and a drainage plane for a given exterior wall system or assembly. See Table 7.5 for a summary of wall system elements.

GUIDELINES FOR SUSTAINABLE CLADDING SELECTION

- Choose a system that will be serviceable at least through the end of the first mortgage term, usually 30 years.
- Compare life-cycle costs for siding, including first cost, maintenance costs, and replacement costs to determine best selection.

TABLE 7.5   WALL SYSTEM ELEMENTS

- Exterior Cladding: The aesthetic face
- Drainage Plane: Bulk water barrier
- Air Barrier System: Infiltration barrier
- Vapor Retarder: Moisture barrier
- Insulating Element: Conduction barrier
- Structural Element: Cladding support

- Use products that come from sustainable sources or those made with recycled or waste material to reduce overall environmental impacts.
- Use products from sustainable manufacturers that can be recycled.[7]

Roofing Materials and Green Roofs

As the use of cool and sustainable roofing systems continues to gain traction in the commercial roofing arena, they are also driving considerable changes in roof system design, roof selection preferences, building codes, and legislation. One result of this change is a growing demand for a new type of roofing system called high-performance roofing that can satisfy cool and sustainable criteria, as well as more established criteria, such as durability, reliability, and cost-effectiveness.

High-performance roofing can contribute to a better indoor environment and improved productivity.

High-performance roofing is part of a larger trend toward constructing high-performance buildings. The U.S. Department of Energy (DOE) has established a high-performance building initiative that promotes energy efficiency nationwide. DOE defines the benefits and objectives of high-performance buildings and whole-building design as:

- Energy consumption reductions of 50 percent or more
- Reduced maintenance and capital costs
- Reduced environmental impact
- Increased occupant comfort and health
- Increased employee productivity

TYPES OF ROOFS

Green roofs

Green roofs, also called living roofs or eco-roofs, are complete roof systems of vegetation, soil, drainage, and a waterproof membrane. They can range from very lightweight systems with as little as 1.5 inches (3.8 centimeters) of soil and low succulent plantings, to deeper heavier systems that support a wide variety of plantings. Because of their water-holding capacity and the presence of vegetation, these roofs have lower solar absorbance properties than most conventional roofs, which essentially keep the roof cooler, resulting in lower energy bills.

The benefits associated with green roofs include the following:

- Control stormwater runoff
- Improve water quality
- Mitigate urban heat island effects
- Filter pollutants and carbon dioxide out of the air
- Prolong the life span of roofing materials

- Conserve energy (reduce building heating and cooling loads)
- Reduce sound reflection and transmission
- Provide a habitat for plants, insects, and wildlife
- Provide additional outdoor useable space for building occupants
- Improve the esthetic environment in both work and home settings

Metal roofs   Metal roofs are usually fabricated from steel or aluminum. They can have a solar reflectance of roughly 60 percent, but they tend to have low emittance. This results in very high surface temperatures and increased energy costs for the facility. Metal roofs are available with a pigmented polymeric coating, similar to paint, that is factory applied. This coating increases the emittance of the roof and produces a solar reflectance nearly as high as the thicker white coatings applied on-site.

Roofing tiles   The typical tiles are either ceramic or cement concrete and are available in a wide range of colors. In general, tile roofs tend to be cooler because of the enhanced air circulation inherent in their installation. Reflectance values for roofing tiles range from 18 to 74 percent, depending on the color.

Reflective roof coatings   These coatings are applied on-site in thicknesses considerably greater than typical white paints, ranging up to about 0.4 inches (1 millimeter). See additional information on roof reflectance and cool roofs later in this section. Types of reflective roof coatings are as follows:

- **White:** White is typically the most reflective of the cool roofs, reflecting up to 70 to 80 percent of the sun's energy.
- **Tinted:** Tinted coatings are usually produced by adding tints to white coatings. This can greatly reduce the solar reflectance, depending on the color. Reflectance values range from 12 to 79 percent.
- **Aluminum:** This type of coating generally employs an asphalt-type resin containing "leafing" aluminum flakes. The aluminum flakes greatly enhance the solar reflectance over the 4 percent value for bare asphalt, to above 50 percent for the most reflective coatings.

Roofing membranes   These are fabricated from strong, flexible, waterproof materials. The upper surface of the membrane may be coated with a pigmented material that determines the color and solar reflectance, or it may simply be topped with roofing gravel. If roofing gravel is used, the membrane has the appearance and solar reflectance of asphalt shingles.

Roof shading   Shading and evapotranspiration (the process by which a plant actively moves and releases water vapor) from trees can reduce surrounding air temperatures as much as 9°F (5°C). Because cool air settles near the ground, air temperatures directly under trees can be as much as 25°F (14°C) cooler than air temperatures above nearby blacktop.

ROOF INSULATION

State and local building codes have requirements for insulation thermal resistances. 47 states, as well as Washington, D.C., have adopted the International Building Code. Chapter 13 of this code (Energy Efficiency), in turn, references the International Energy Conservation Code (IECC). The IECC has adopted American Society of Heating, Refrigerating, and Air-Conditioning Engineers (ASHRAE) 90.1-2004, *Energy Standard for Buildings Except Low-Rise Residential Buildings,* as a reference standard. The ASHRAE 90.1 standard separates the United States into eight climate zones, with Zone 1 being the hottest and Zone 8 being the coldest.

Based on the climate zones, the standard also lists the recommended thermal resistance for continuous insulation located above the roof deck.

A basic understanding of available insulation types and the code requirements for insulation thermal resistance is essential before selecting the type of insulation to use under a roof covering. Listed below are approximate material thicknesses required to achieve a thermal resistance of R-20:

- Wood fiber = 7.5 inches (19.1 centimeters)
- Perlite = 7.5 inches (19.1 centimeters)
- Polyisocyanurate = 3.3 inches (8.4 centimeters)
- Expanded polystyrene = 5.3 inches (13.5 centimeters)
- Extruded polystyrene = 4 inches (10.2 centimeters)
- Cellular glass = 4.5 inches (11.4 centimeters)
- Composite (polyisocyanurate/wood fiber, perlite, or gypsum board) = 3.5 inches (8.9 centimeters)

Selecting a roof insulation product cannot be isolated because the choice generally needs to be made in close connection with the membrane type and application method. Consider selecting an insulation product that includes these attributes:

- Membrane compatibility
- Energy code (R-value requirements)
- Insurance and code requirements regarding wind and fire
- Strength (more specifically, compressive strength for traffic and hail resistance)

Not all insulation types are compatible with all roof covering types. For instance, expanded and extruded polystyrenes should not be in direct contact with thermoplastic polyvinyl chloride (PVC) single-ply membranes. When in direct contact, plasticizer migration occurs, causing the PVC membrane to become brittle. Modified bitumen membranes, as well as conventional built-up membranes, are not recommended for direct application over foam plastic insulation boards. Cover board insulation is recommended and required by modified bitumen and built-up roof covering manufacturers when utilizing foam plastic insulations. Wood fiber, perlite, and gypsum board are the most common cover board insulations in use today.

ROOF REFLECTANCE (COOL ROOFS)

The ideal cool roof is one whose surface is minimally heated by the sun, such as a bright white roof. However, sometimes the word *cool* is more appropriately used to describe a roofing product whose surface is warmer than that of a bright white material, but still cooler than that of a comparable standard product. For example, the afternoon surface temperature of a specially designed cool red roof is higher than that of a bright white roof, but lower than that of a standard red roof. An example of a hot roof is one with a standard black surface, which grows very warm in the sun.[8]

On a sunny day, roof temperatures can range from comfortably warm to egg-frying hot, depending on how much sunlight they reflect. Different roofing materials were tested side-by-side by Lawrence Berkeley Laboratory researchers. The peak temperatures of various roofing colors are listed below. Ambient air temperature at the time of the test was 55°F (13°C):

- Black acrylic paint: 142°F (61°C)
- Galvanized steel: 138°F (59°C)
- Black acrylic paint infrared-reflecting film: 123°F (51°C)
- Common white fiberglass/asphalt shingle: 118°F (48°C)
- Clay terra-cotta tile: 112°F (44°C)
- Red acrylic paint: 106°F (41°C)
- Light green acrylic paint: 104°F (40°C)
- White acrylic paint: 74°F (23°C)
- Hyper-white acrylic paint: 65°F (18°C)

The most important feature of an ideal cool roof is that its surface strongly reflects sunlight. The surface of an ideal cool roof should also efficiently cool itself by emitting thermal radiation. Thus, a cool roof should have both high "solar reflectance" (ability to reflect sunlight, measured on a scale of 0 to 1) and high "thermal emittance" (ability to emit thermal radiation, also measured on a scale of 0 to 1). The solar reflectance and thermal emittance of a surface are called its radiative properties because they describe its abilities to reflect solar radiation and emit thermal radiation. On a clear day, about 80 percent of sunlight reflected from a horizontal roof will pass into space without warming the atmosphere or returning to the earth.

There are a variety of roofing assembly technologies that can reduce the flow of heat into a building, including solar-reflective, thermally emissive surfaces, vegetative cover (green roofing), thermally mass construction, super-insulation, and ventilation. However, we reserve the term *cool roof* to refer to one that stays cool in the sun by virtue of high solar reflectance and high thermal emittance.

Reducing the building peak cooling load with a cool roof can allow the installation of a smaller, less expensive air conditioner. This is referred to as a "cooling equipment" saving. Smaller air conditioners are also typically less expensive to run, because air conditioners are more efficient near full load than at part load. Choosing a cool roof instead of a standard roof can slightly increase the need for heating energy in winter.

However, winter penalties are often much smaller than summer savings even in cold climates because the northern mainland United States (latitude ≥ 40°N) receives about three to five times as much daily sunlight in summer as in winter.

Certain government agencies and industry groups have established guidelines for the testing and performance of cool roofing[9]:

- The ENERGY STAR Reflective Roof Products Program has established a minimum standard that requires low-slope roof products to have an initial reflectance of at least 65 percent and a reflectance of at least 50 percent after three years of weathering. If there is any doubt about whether a roofing system is "cool" or energy efficient, check to see if it is listed in the ENERGY STAR Roof Products listings at www .energystar.gov.
- The American Society of Heating, Refrigeration, and Air Conditioning Engineers established Standard 90.1 to set minimum requirements for energy-efficient building design. Adopted by the federal government in 1994, it sets the roof reflectance minimum for government facilities at 70 percent and the minimum emittance level at 75 percent.
- The Cool Roof Rating Council (CRRC), Oakland, California, is a nonprofit association that implements and promotes fair, accurate performance ratings for solar reflectance and emittance from roof surfaces. CRRC's Product Rating Program enables roofing manufacturers to label various roof-surface products with radiative property values rated under strict guidelines. Performance data for products from numerous manufacturers can be found on the CRRC website at www.coolroofs.org.

ACHIEVING A SUSTAINABLE ROOF

In its primer on Sustainable Building Design, the Rocky Mountain Institute in Snowmass, Colorado, maintains that sustainable roofing can be accomplished in five different ways:

- Amount of recycled content in the roofing product
- The use of recyclable materials, such as thermoplastics and metal
- An extended roof service life
- More efficient use of energy and other natural resources
- The actual renewal of natural resources through using recycled content in new roofing

One aspect of achieving roofing sustainability is reducing waste when reroofing or installing a new roof on a building for the first time. While roofing waste cannot be eliminated altogether, there are ways to significantly reduce the amount of waste[10]:

- Building owners can consider installing roofing systems directly over existing asphalt, metal, and/or single-ply systems, depending on the level of saturation and certain other conditions. In many cases, thermoplastic, single-ply systems can be installed directly over existing roofs and still be fully warranted.

- Building owners can choose single-ply systems that are custom prefabricated to fit each building. Custom prefabrication significantly reduces installation waste, requiring many fewer trips to the landfill, regardless of the size of the project.
- Building owners can choose a roofing system that can be recycled after its useful life to make products such as flooring, park benches, or new roofing components. Metal and thermoplastic single-ply materials, such as PVC and thermoplastic polyolefin, are recyclable in the plant, after installation, and after the roof's useful service. Some manufacturers of metal and thermoplastic roofing systems have initiated recycling programs.
- New technologies have made it possible for thermoset ethylene propylene diene monomer (EPDM) single-ply components to be recycled into materials for lower level use, though the cost can be high.

Durability/life span   Many factors affecting the durability and longevity of roofing systems are beyond an owner's control. These factors include climate, catastrophic accidents, and violent storms. In terms of high-performance roofing, durability is the ultimate reflection of the performance of every roofing component or element that can be controlled by intelligent design, manufacture, installation, and maintenance[11]:

- Proper design for:
  - Location
  - Climate
  - Roof deck/substrate
  - Building occupancy
- Roof deck matched to the insulation/membrane systems
- Proper drainage to prevent water ponding
- Quality of roofing product and manufacture:
  - Tensile strength
  - Water absorption
  - Fire resistance
  - Wind uplift
  - Elongation and thermal expansion
  - Dynamic puncture resistance
  - Resistance to rooftop contaminants
  - Resistance to solar degradation
  - Flashing details at roof penetrations and walls
- Quality of installation and maintenance:
  - Regular curb inspection and maintenance
  - Roof traffic tiles
  - Limited and supervised roof traffic
  - Periodic roof inspections and spot repairs by roofing contractor
  - Single warranty for entire roof system

See Table 7.6 for a summary of sustainable roof attributes.

TABLE 7.6   SUSTAINABLE ROOF ATTRIBUTES

- Proper design for the climate and locale
- Deck, insulation, and membrane designed as a system
- Drainage design to prevent ponding
- Selection of quality product and installer
- Single system warranty; not component warranty
- Roof system has recycled content
- Roof system is recyclable
- Periodic inspections and maintenance of roof and flashing
- Limited and controlled roof traffic

ROOF SUSTAINABILITY RECOMMENDATIONS

The following roof systems are recommended, assuming compliance with sections above. Note: Special site or building conditions may preempt the following recommendations.[12]

Roofs with slopes less than 1:12

1 Metal panel systems are not recommended.
2 Protected membrane roofs are recommended for low-slope situations. For the membrane, modified bitumen is the ideal product from an energy-efficiency point of view (though not necessarily from a sustainability viewpoint). If there is compelling reason to specify a single-ply membrane, EPDM is a most desirable product. For ballast, concrete pavers or Extruded expanded polystyrene (XEPS) boards with a factory-applied mortar surface are recommended in lieu of aggregate.
3 When using a modified bitumen membrane roof, attach the membrane with cold adhesive or use a torched application (where torching is allowed).

Roofs with slopes greater than 1:12, and less than or equal to 3:12

For roofs of moderate slope, standing seam hydrostatic metal panel systems (as well as all of the low-slope systems) are the most desirable option due to their durability, sustainability, and energy-efficiency characteristics.

Roof with slopes greater than 3:12

At slopes greater than 3:12, aesthetics often play a significant role in system selection. Selecting a roofing system primarily based on aesthetics is acceptable, provided the roof has an acceptable balance of energy-efficiency, durability, and sustainable attributes. Roof types appropriate for slopes greater than 3:12 are as follows:

1 Standing seam hydrostatic or hydrokinetic panels.
2 Asphalt shingles are typically specified where initial cost is a primary factor in system selection.

3 Copper panel systems, slate, and tile are typically specified where there is compelling aesthetic reason to specify these relatively expensive systems.

4 Modified bitumen, single-ply, and spray polyurethane foam roofing systems may be suitable on steep-slope roofs, provided snowslides are not problematic. If modified bitumen or single-ply membranes are specified, be aware of aesthetic issues (e.g., seam lines will be visible and highly reflective surfaces may become discolored and visually objectionable).

Site Lighting

Site lighting should direct the appropriate amount of light on illuminated surfaces without contributing to glare and light pollution. Site lighting should be sensitive to the natural environment and should not disturb the nocturnal habits of wildlife. For instance, lighting levels in rural areas and parks should be lower than those in suburban and urban areas. In the past, general illumination levels were high, but with a better understanding of lighting design, general or ambient lighting levels are now lower, with more emphasis on lighting the task, whether it be assembling electronic components, reading a book, or working at a desk in front of a computer.

Efficient site lighting systems shine light on the area that needs to be lit, without wasting energy or producing unusable light and glare. This is accomplished by using full cutoff luminaires that do not shine any light above the horizon line. Shorter poles and landscape lights produce light nearer to the ground, thus minimizing glare and light pollution. Tall lighting poles are to be used primarily for parking lots where multiple luminaries are mounted on a single pole to reduce the number of poles needed. Typically, full cutoff luminaires use flat glass lenses rather than dropped refractors. Efficient site lighting systems must provide uniform illumination levels so that steps, curbs, and obstacles are not hidden in shadows. Design lighting controls into the project to turn off exterior lighting when the area is unoccupied overnight or after closing time.

ASHRAE/Illuminating Engineering Society of North America (IESNA) 90.1-2004 provides requirements for exterior lighting, including facades, architectural features, entrances, exits, loading docks, and illuminated canopies, as well as exterior building grounds provided through the building's electrical service. The standard also has provisions for exterior lighting controls to turn off exterior lighting when sufficient daylight is available or when the lighting is not required during nighttime hours.

The U.S. Green Building Council's LEED rating system includes credit for eliminating light spillage from the building and site. The credit includes the requirement that exterior luminaires with more than 1000 initial lamp lumens are shielded, and all luminaires with more than 3500 initial lamp lumens meet the full cutoff IESNA Classification. LEED Sustainable Site Credit 8 has specific requirements to be met based on the specific zone, from wilderness and state parks to city centers. The requirement is not just for parking lots; it is also for landscape lighting and building facades. The GreenGlobes rating system from the Green Building Initiative also has points under site design related to site development for minimizing the obtrusive aspects of exterior lighting, including glare, light trespass, and sky glow.

The use of lighting controls for exterior lighting is gaining acceptance. For buildings that are normally only occupied in the daytime, a significant amount of energy is wasted by lighting the parking lot all night. For safety reasons, parking lots should not be totally dark, but should maintain a low level of illumination. Also, high-intensity discharge lighting must have a warm-up time of several minutes. For this reason, they cannot be turned off completely as a part of an energy-saving strategy. Lighting controls are now available that sense when someone enters the parking lot and raise the lighting level to full brightness for a period of time. If no motion is detected after a preset time, the parking lot lights will return to the lower light level. Lowering site lighting levels reduces light pollution and minimizes energy consumption.[13]

Site Design

Decreasing impervious surfaces on a site is the most basic and simple strategy to address water quality concerns and avoid a host of site-related sustainability problems, including stormwater runoff and water table depletion.

IMPERVIOUS AREA REDUCTION

Methods of reducing the impervious surface area include the following:

- Reduction of roadway surfaces can retain more permeable land area. In some cases, planners have reduced pavement needs by up to 40 percent by using longer, undulating roads that create more available lot frontage instead of the more common alternative of designing with wider, shorter streets with more intersections. Other options to reduce the paving area may include shared driveways, zero-lot line lots with reduced street frontage, landscaped detention islands within cul-de-sacs, or alternate designs for turnaround areas.
- Permeable pavement surfaces can be constructed from a variety of materials, including traditional asphalt and concrete, gravel, or pavers. Permeable roads and parking areas allow water to flow through them, reducing runoff and replenishing the soil areas directly beneath the paving. However, the subbase underneath permeable pavements must be engineered to accommodate both filtration and a sufficient quantity of water storage. In many cases, permeable surfaces can reduce or eliminate the need for traditional stormwater structures.
- Vegetative roof systems create a lightweight, permeable vegetative surface on an impervious roof area. Moss, grass, herbs, wildflowers, and native plants can be used, creating an aesthetically pleasing roof landscape. The systems start with a high-strength rubber membrane placed over the base roof structure. Various layers above the rubber may contain insulation, filter and drainage media, separation fabrics,

lightweight growth media, vegetation, and wind erosion fabric. Some systems even incorporate rain barrel runoff collection, pumping, and irrigation equipment. These systems are more costly than standard roofs, and have not been used on a large scale for residential development in the United States. See Fig. 7.2 for a photograph of a vegetative roof system.

■ Planning site layout and grading to natural land contours can minimize grading costs and retain a greater percentage of the land's natural hydrology. Contours that function as filtration basins can be retained or enhanced for water quality and quantity, and incorporated into the landscaping design.

■ Natural resource preservation and xeriscaping can be used to minimize the need for irrigation systems and enhance property values. Riparian, or stream bank, areas are particularly crucial to water quality, and in most areas, subject to Federal or State regulations. Preservation of existing wooded areas, mature trees, and natural terrain can give new developments a premium mature landscape appearance and provide residents with additional recreational amenities. Both of these features can improve marketability. Xeriscaping refers to landscaping with plants native to area climate and soil conditions. These plants thrive naturally, requiring less maintenance and irrigation than most hybrid or imported varieties.

LOW-IMPACT DESIGN STRATEGIES

Low-impact strategies can be used to address water quality issues that fall under two broad categories called practices and site design. Water quantity management for storm

Figure 7.2   **Vegetative roof gardens create more permeable and cooler roofs.**

events may be required in the stormwater design as well. The most common concepts are summarized below:

Practices   The basic strategy for handling runoff is to either reduce the volume of runoff or decentralize the runoff flow. Both of these goals are best accomplished by creating a series of smaller retention/detention areas that allow localized filtration (rather than carrying runoff to a remote collection area), in conjunction with facilities addressing larger storm events as required. For the practices noted below, special attention should be paid to the composition of existing soils, as well as new soils or amended soils used, and underlying topography. For instance, a locale with eroded and pitted topography may react differently to the introduction of low-impact practices than a site that does not have similar ground openings and underground channels. Common methods of reducing/decentralizing the stormwater runoff on a site include:

- Bioretention cells typically consist of grass buffers, sand beds, a ponding area for excess runoff storage, organic layers, planting soil, and vegetation. Their purpose is to provide a storage area, away from buildings and roadways, where stormwater collects and filters into the soil. Permanent ponds can be incorporated into the cell design as quantity management and landscaping features. Temporary storage areas without ponds may be called detention cells. Bioretention areas have also been called rain gardens since they are typically landscaped with native plants and grasses, selected according to their moisture requirements and ability to tolerate pollutants. Annual maintenance of bioretention cells must be planned in order to replace mulching materials, remove accumulated silt, or revitalize soils as required.
- Vegetated swales can be used as alternatives to curb and gutter systems, and are usually installed along residential streets or highways. They use grasses or other vegetation to reduce runoff velocity and allow filtration, while high-volume flows are channeled away safely to a quantity management facility. Features like plantings and check dams may be incorporated to further reduce water velocity and encourage filtration. Walkways are either separated from roadways by swales, or relocated to other areas. In areas where salts are commonly used for winter deicing, careful attention must be paid to selecting plant species that are salt tolerant.
- Filter strips can be designed as landscape features within parking lots or other areas, to collect flow from large impervious surfaces. They may direct water into vegetated quantity detention areas or special sand filters that capture pollutants, and gradually discharge water over a period of time.
- Disconnected impervious areas direct water flows collected from structures, driveways, or street sections into separate localized detention cells instead of combining it in drainpipes with other runoff. Disconnecting the flow limits the velocity and overall amount of conveyed water that must be handled by end-of-pipe water quality and quantity facilities.
- Cistern collection systems can be designed to store rainwater for dry-period irrigation, rather than channeling it to streams. Smaller tanks that collect residential roof drainage are often called "rain barrels" and may be installed by individual

homeowners. Some collection systems are designed to be installed directly under permeable pavement areas, allowing maximum water storage capacity while eliminating the need for gravel beds. Other innovative systems incorporate gray water collection for additional water conservation.

COST BENEFITS

The cost benefits to builders and developers utilizing low-impact design strategies can be significant. According to the Center for Watershed Protection, traditional curbs, gutters, storm drain inlets, piping, and detention basins can cost two to three times more than engineered grass swales and other techniques to handle roadway runoff. Other strategies can have similar impact. Choosing permeable pavement for a parking area may eliminate the need for a catch basin and conveyance piping to service it. Small distributed filtration areas on individual lots can reduce site requirements for larger detention ponds (perhaps with associated headwalls, fencing, and bank control) that take up valuable land area.

Not all sites can effectively utilize low-impact design techniques. Soil permeability, slope, and water table characteristics may limit the potential for local infiltration. Urban areas and locations with existing high contaminant levels may be precluded from using filtration techniques.

Many existing local codes, zoning regulations, parking requirements, and street standards were developed prior to the emergence of water quality and stormwater management concerns, and may prohibit or inhibit implementing low-impact practices.[14]

Heat Island Reduction

The term *heat island* describes built-up areas that are hotter than nearby rural areas. The annual mean air temperature of a city with 1 million people or more can be 1.8–5.4°F (1–3°C) warmer than its surroundings. In the evening, the difference can be as high as 22°F (12°C). Heat islands can affect communities by increasing summertime peak energy demand, air conditioning costs, air pollution and greenhouse gas emissions, heat-related illness and mortality, and water quality.

Elevated temperature from urban heat islands, particularly during the summer, can affect a community's environment and quality of life. While some heat island impacts seem positive, such as lengthening the plant-growing season, most impacts are negative and include[15]:

■ **Increased energy consumption:** Higher temperatures in summer increase energy demand for cooling and add pressure to the electricity grid during peak periods of demand. One study estimates that the heat island effect is responsible for 5–10 percent of peak electricity demand for cooling buildings in cities.

■ **Elevated emissions of air pollutants and greenhouse gases:** Increasing energy demand generally results in greater emissions of air pollutants and greenhouse gas emissions from power plants. Higher air temperatures also promote the formation of ground-level ozone.

- **Compromised human health and comfort:** Warmer days and nights, along with higher air pollution levels, can contribute to general discomfort, respiratory difficulties, heat cramps and exhaustion, nonfatal heat stroke, and heat-related mortality.
- **Impaired water quality:** Hot pavement and rooftop surfaces transfer their excess heat to stormwater, which then drains into storm sewers and raises water temperatures as it is released into streams, rivers, ponds, and lakes. Rapid temperature changes can be stressful to aquatic ecosystems.

SITE STRATEGIES TO REDUCE HEAT ISLANDS

- Trees and vegetation
- Green roofs
- Cool roofs
- Cool pavements
- Plant trees and vegetation

Trees and vegetation are most useful as a mitigation strategy when planted in strategic locations around buildings or to shade pavement in parking lots and on streets. Researchers have found that planting deciduous trees or vines to the west is typically most effective for cooling a building, especially if they shade windows and part of the building's roof.

A green roof, or rooftop garden, is a vegetative layer grown on a rooftop. Green roofs provide shade and remove heat from the air through evapotranspiration, reducing temperatures of the roof surface and the surrounding air. On hot summer days, the surface temperature of a green roof can be cooler than the air temperature, whereas the surface of a conventional rooftop can be up to 90°F (50°C) warmer. Green roofs can be installed on a wide range of buildings, from industrial facilities to private residences. They can be as simple as a 2-inch (5.1-centimeter) covering of hardy ground cover or as complex as a fully accessible park complete with trees. Green roofs are becoming popular in the United States, with roughly 8.5 million square feet (79 hectares) installed or in progress as of June 2008.

Cool roofs   A high solar reflectance or albedo is the most important characteristic of a cool roof as it helps to reflect sunlight and heat away from a building, reducing roof temperatures. A high thermal emittance also plays a role, particularly in climates that are warm and sunny. Together, these properties help roofs to absorb less heat and stay up to 50–60°F (28–33°C) cooler than conventional materials during peak summer weather.

Based on 2006 data from more than 150 ENERGY STAR partners, shipments of cool roof products have grown to represent more than 25 percent of these manufacturers' commercial roof products.

Cool pavements   Cool pavements include a range of established and emerging technologies that communities are exploring as part of their heat island reduction efforts. The term currently refers to paving materials that reflect more solar energy, enhance water evaporation, or have been otherwise modified to remain cooler than conventional pavements.

Conventional paving materials can reach peak summertime temperatures of 120–150°F (48–67°C), transferring excess heat to the air above them and heating stormwater as it runs off the pavement into local waterways. Due to the large area covered by pavements in urban areas (nearly 30–45 percent of land cover based on an analysis of four geographically diverse cities), they are an important element to consider in heat island mitigation.

Cool pavements can be created with existing paving technologies (such as asphalt and concrete) as well as newer approaches, such as the use of coatings or grass paving. Cool pavement technologies are not as advanced as other heat island mitigation strategies, and there is no official standard or labeling program to designate cool paving materials.

Parking structures   With some occupancies, parking structures may occupy as much site area as the building footprint. Since the concrete construction of parking decks also provides a considerable amount of mass storage that contributes to the heat island effect, it is important to use light-colored concrete on the upper level of parking structures to help reduce the absorption of energy into the structure.

Site Disturbance

The goal of minimizing site disturbance during and after construction is for the purpose of:

1 Conserving existing natural areas
2 Restoring damaged areas
3 Providing habitat and promoting biodiversity

Site development strategies are implemented by carefully siting the building to minimize disturbance to existing ecosystems, restoring damaged areas, and designing the building to minimize its footprint.

Trees can be protected during construction by avoiding compaction and disturbance of soil around the building at the job site, especially soil over roots. In addition, the area under tree drip lines should be fenced and made off-limits to all foot and vehicular traffic, and tree trunks and exposed branches should be well protected with strong wood barriers. Arranging truck access to minimize the need for trucks to reposition or turn around will also help protect the site.

LOW-IMPACT DEVELOPMENT

Use Low-Impact Development strategies to design and locate the site to preserve the natural environment and reduce stormwater runoff. By planning for terrain, vegetation, and soil features that handle stormwater on site, you can avoid costly storm drain systems and water treatment. Limit impervious areas or use permeable pavement or paved surfaces that direct runoff into bioretention gardens and vegetative swales. Plan to save natural vegetative areas and large trees, and have a qualified person on-site to see that these measures are implemented. Minimize slope and site disturbance, reestablish ground cover within 14 days of disturbance, and retain the natural topography, flora, and fauna of the site.

Endnotes

1. Sustainable Landscape Council. "Standards & Guidelines." http://www.sustainable-landscapecouncil.com/. Accessed March 10, 2010.
2. Montgomery County (PA) Planning Commission. "Planning by Design." http://planning.montcopa.org/planning/cwp/fileserver,Path,PLANNING/Admin%20-%20Publications/Planning%20by%20Design/Sustainable_Paving_web.pdf,assetguid,2e12932a-a896-420c-97184af6ed56f0a9.pdf. Accessed April 12, 2010.
3. Water Environment Research Foundation. "Basic Principles." http://www.werf.org/livablecommunities/pdf/basic.pdf. Accessed February 18, 2010.
4. Lawrence Berkeley National Laboratory. "Building Shading Strategies." http://windows.lbl.gov/pub/designguide/section5.pdf. Accessed April 28, 2010.
5. National Institute of Building Sciences Whole Building Design Guide. "Sun Control & Shading Devices." http://www.wbdg.org/resources/suncontrol.php. Accessed February 18, 2010.
6. Regents of the University of Minnesota. "Minnesota Green Affordable Housing Guide." http://www.greenhousing.umn.edu/comp_cladding.html. Accessed April 28, 2010.
7. National Institute of Building Sciences Whole Building Design Guide. "Building Envelope Design Guide-Wall Systems." http://www.wbdg.org/design/env_wall.php. Accessed April 28, 2010.
8. Levinson, Ronnen: Lawrence Berkeley Laboratory. "Cool Roof Q&A." http://coolcolors.lbl.gov/assets/docs/fact-sheets/Cool-roof-Q%2BA.pdf. Accessed August 12, 2010.
9. Woolridge, Mike: Lawrence Berkeley Laboratory. "Cool ideas for roofs cut energy bills, smog." http://www.lbl.gov/Science-Articles/Archive/heat-islands-and-roof-color.html. Accessed September 10, 2010.
10. Ballensky, Drew: Commercial Building Products magazine. "Five Factors Drive Roof Sustainability." http://www.cbpmagazine.com/article.php?articleid=140. Accessed September 15, 2010.
11. Nunnikhoven, Alvin: Buildings. com. "Tips for Selecting Rigid Roof Insulation." http://www.buildings.com/ArticleDetails/tabid/3321/ArticleID/5646/Default.aspx. Accessed August 12, 2010.
12. National Institute of Building Sciences Whole Building Design Guide. "Building Design Envelope Guide-Roofing Systems." http://www.wbdg.org/design/env_roofing.php. Accessed April 28, 2010.

13. American Institute of Architects. "Efficient Artificial and Site Lighting." http://wiki .aia.org/Wiki%20Pages/Efficient%20Artificial%20and%20Site%20Lighting.aspx. Accessed August 12, 2010.

14. National Association of Homebuilders Toolbase Services. http://www.toolbase .org/Building-Systems/Sitework/low-impact-development. Accessed February 20, 2010.

15. U.S. Environmental Protection Agency. "Heat Island Effect." http://www.epa.gov/ heatisland/about/index.htm. Accessed August 12, 2010.

8

RESOURCES

Manufacturers

ENERGY SERVICES COMPANIES

National Association of Energy Service Companies: http://www.naesco.org

- AECOM Energy, San Diego, CA
- AMERESCO, Framingham, MA
- Burns & McDonnell, Kansas City, MO
- CCI Group, LLC, Shalimar, FL
- Chevron Energy Solutions, Overland Park, KS
- Clark Energy Group, LLC, Arlington, VA
- CM3 Building Solutions, Inc., Trevose, PA
- Comfort Systems USA Energy Services, Windsor, CT
- ConEdison Solutions, Valhalla, NY
- Constellation Energy Projects & Services Group, Pittsburgh, PA
- Control Technology and Solutions (CTS), St. Louis, MO
- Eaton Corporation, Raleigh, NC
- Energy Focus, Inc., Milwaukee, WI
- Energy Solutions Professionals, LLC, Overland Park, KS
- Energy Systems Group, Newburgh, IN
- FPL Energy Services, West Palm Beach, FL
- Honeywell International, Inc., Danvers, MA
- Johnson Controls, Inc., Milwaukee, WI
- Lockheed Martin, Arlington, VA
- McClure Company, Harrisburg, PA
- NORESCO, Westborough, MA
- Onsite Energy Corporation, Carlsbad, CA
- Pepco Energy Services, Inc., Arlington, VA
- Schneider Electric, Harrisburg, PA

■ Siemens Industry, Inc., Buffalo Grove, IL
■ Synergy Companies, Orem, UT
■ The EnergySolve Companies, Somerset, NJ
■ Trane, St. Paul, MN
■ UCONS, LLC, Kirkland, WA
■ Wendel Energy Services, Amherst, NY

Incentives and Resources

■ The U.S. Department of Energy (Energy Efficiency and Renewable Energy Program) maintains an online database of financial incentives and grants targeted to business, industry, and universities at: http://www1.eere.energy.gov/financing/.
■ Current U.S. federal government business tax incentives for energy-efficiency improvements can be viewed at: http://www1.eere.energy.gov/buildings/tax_commercial.html.
■ U.S. consumer tax credits (which include the hybrid gas/electric and alternative fuels tax credits for businesses) can be viewed at: http://www.energy.gov/taxbreaks.htm.
■ Energy-savings investment tax breaks for businesses can be viewed at: http://www.energy .gov/additionaltaxbreaks.htm.

Software and Calculators

DISK-BASED TOOLS

■ Environmental Energy Technologies Division (EETD) Software: http://eetd.lbl.gov/ eetd-software.html
A complete listing of Division software.
■ U. S. Department of Energy Building Tools Directory: http://www.eere.energy.gov/ buildings/tools_directory/
This is an electronic directory providing information about the range of buildings-related energy tools available to the buildings industry. These tools include databases, spreadsheets, component and system analysis, and whole-building energy performance simulations programs.
■ CalArch: http://eetd.lbl.gov/eetd-software-calarch.html
CalArch is a web-based tool for benchmarking the whole building energy use of California commercial buildings.
■ COMIS: http://eetd.lbl.gov/eetd-software-comis.html
COMIS is a computer model that simulates the air flow distribution in multizone buildings. This program was developed in an international effort by researchers from nine countries.
■ DER-CAM: http://eetd.lbl.gov/eetd-software-der-cam.html
The Distributed Energy Resources Customer Adoption Model (DER-CAM) is a sophisticated economic model to track customer DER adoption.

■ Design Intent Tool: eetd-software-dit.html
This tool helps building owners, architects, and engineers develop a Design Intent Document to facilitate record-keeping, and ensure that the owner's and designer's goals are achieved and periodically verified through building performance measurement.

■ DOE-2: http://eetd.lbl.gov/eetd-software-doe2.html
It is an advanced computer program that simulates hourly building energy use.

■ EnergyPlus: http://eetd.lbl.gov/eetd-software-eplus.html
This program takes the best features of the DOE-2 and BLAST programs, and combines them into a more powerful and accurate program.

■ GenOpt: http://eetd.lbl.gov/eetd-software-genopt.html
It is an optimization program that allows the use of a number of inputs. The program is designed to assess the best ways to minimize annual energy use that is calculated by an external simulation program, such as EnergyPlus or DOE-2.

■ Home Energy Saver: http://eetd.lbl.gov/eetd-software-hes.html
It is the first Internet-based tool for calculating energy use in residential buildings. Though Intended mainly for homes, this tool can be useful in calculating the performance of small commercial and retail facilities as well. The project was sponsored by the U.S. Environmental Protection Agency and the U.S. Department of Energy as part of their national ENERGY STAR® Programs for improving energy efficiency in homes.

■ Interoperability (IAI): http://eetd.lbl.gov/eetd-software-iai.html
The International Alliance for Interoperability's goal is to develop a standard universal framework to enable information sharing across all phases of the building life cycle. This may be an increasingly popular tool as IgCC is adopted and more building life-cycle data is collected.

■ Metracker: http://eetd.lbl.gov/eetd-software-metracker.html
Metracker is a prototype tool designed to allow the specification, tracking, and charting of building performance objectives and their associated metrics across the complete life cycle of a building.

■ Optics: http://eetd.lbl.gov/eetd-software-optics.html
Optics is a computer program designed to calculate the optical properties of glazing systems and laminates. The program can be used to construct new laminates from existing components and manipulate glazing systems by adding coatings and films and modifying the substrate or thickness. The program incorporates the International Glazing Database, which contains spectral data of more than 1000 different types of glazings.

■ PEAR: http://eetd.lbl.gov/eetd-software-pear.html
PEAR is a simplified tool based on extensive DOE-2 simulations. It is readily useable by builders, architects, and lenders to provide reliable estimates of building energy consumption. The software can also be used to produce residential energy ratings.

■ ProForm: http://eetd.lbl.gov/eetd-software-proform.html
ProForm is a software tool designed to perform a basic assessment of the environmental and financial impacts of renewable energy and energy-efficiency projects. ProForm provides a basic financial measurement and calculates the reduced or avoided local air pollutants and emissions of CO_2 from a project.

- RADIANCE: http://eetd.lbl.gov/eetd-software-rad.html
  Radiance is a photometrically accurate graphic simulation of the effects of lighting in indoor environments. The realistic images produced by the program aid the designer in visualizing lighting design options.
- RESEM: http://eetd.lbl.gov/eetd-software-resem.html
  RESEM (Retrofit Energy Savings Estimation Model) calculates long-term energy savings directly from actual utility data, and includes the ability to modify the calculation for weather and building use variations during the renovation or retrofit period.
- RESFEN: http://eetd.lbl.gov/eetd-software-resfen.html
  RESFEN(RESidential FENestration) is a program developed for calculating the annual heating and cooling energy costs due to fenestration/glazing systems in residential buildings. RESFEN also calculates the fenestration contribution to peak heating and cooling loads, and can be used on small commercial and retail buildings as well
- SPARK: http://eetd.lbl.gov/eetd-software-spark.html
  SPARK is software that uses an object-oriented environment to create graphic models of complex commercial building energy systems.
- Superlite: http://eetd.lbl.gov/eetd-software-superlite.html
  It is a sophisticated (mainframe and microcomputer) program that calculates the distributions of daylight illuminance for complex room and light source configurations. Superlite will model daylight entering the space from as many as five openings, and can calculate the benefits of reflected light from as many as 20 opaque surfaces of various orientations.
- THERM: http://eetd.lbl.gov/eetd-software-therm.html
  THERM models two-dimensional heat transfer effects through combinations of building components.
- Window: http://eetd.lbl.gov/eetd-software-window.html
  Window is a thermal analysis program that is used by a wide range of U.S. manufacturers to document their glazing and window product performance.
- U.S. Department of Energy (DOE), Office of Energy Efficiency and Renewable Energy
  Many of the DOE's Office of Energy Efficiency and Renewable Energy programs develop software tools to help researchers, designers, architects, engineers, builders, code officials, and others involved in the building life cycle in evaluating and ranking potential energy-efficiency in new or existing buildings. This electronic directory provides information about the range of buildings-related energy tools available to the buildings industry. Many tools presented in this directory have been or are currently sponsored by DOE. The tools range from tools intended primarily for research to software now available from private sector vendors.

  The energy tools in this directory include databases, spreadsheets, component and systems analyses, and whole-building energy performance simulation programs. For each tool in the directory, a short description is provided along with other information including expertise required, users, audience, input, output, computer platforms, programming language, strengths, weaknesses, technical contact, and availability.

To help you find the tool you are most interested in, the directory is organized in the following categories:

- Whole-Building Energy Analysis
- Material, Component, Equipment, and Systems Evaluation
- Codes and Standards Development and Compliance
- Special Applications

WEB-BASED TOOLS AND CALCULATORS

An increasing number of energy calculators can be found on the Web. Many are limited to a single equipment type, fuel, or locality. For example, a number of energy utilities now offer Web-based calculators that apply to individual fuels or local conditions.

The following list focuses on the more widely applicable calculators available on the U.S. Department of Energy (DOE) Energy Efficiency and Renewable Energy (EERE) website at http://apps1.eere.energy.gov/buildings/tools_directory/.

- Athena Model:
  A comprehensive database for making building life-cycle assessments, and evaluating the environmental characteristics (sustainability) of building materials and components.
- Autodesk Green Building Studio:
  Building information modeling, interoperability, energy performance, DOE-2, EnergyPlus, CAD
- BEES:
  A program for assessing environmental performance, green buildings, life-cycle assessment, life-cycle costing, and sustainable development
- BLCC:
  Performs economic analysis, ESPCs for federal buildings, as well as life-cycle costing
- Building Design Advisor:
  A program that analyzes design, daylighting, energy performance, prototypes, case studies, and commercial buildings
- CATALOGUE:
  Software for calculating the energy performance and thermal characteristics of windows and fenestration products
- Climate Consultant:
  Performs climate analyses; generates psychrometric charts, bioclimatic charts, and wind wheels
- COMcheck:
  Performs basic energy code compliance for commercial buildings; acceptable in many building code jurisdictions
- CONTAM:
  Performs airflow analysis; assesses the benefits of: building controls, contaminant dispersal, indoor air quality, multizone analysis, smoke control, smoke management, and ventilation

- Cool Roof Calculator:
  Assesses reflective roof, roofing membrane, and low-slope roof systems
- CYPE-Building Services:
  Estimates building services, single model, energy simulation, sizing, HVAC, plumbing, sewage, electricity, solar, and analysis of acoustic behavior
- Daylight:
  Calculates daylighting benefits and daylight factors
- DAYSIM:
  Performs annual daylight simulations, electric lighting energy use, lighting controls
- Demand Response Quick Assessment Tool:
  Allows the user to estimate demand response and overall building energy loads
- Design Advisor:
  A whole-building assessment tool, useful for reviewing energy consumption, thermal comfort, natural ventilation, and double-skin facades
- Discount:
  Calculates present value, discount factors, future values, and life-cycle cost
- EE4 CODE:
  A program that includes assessments for standards and code compliance, as well as whole building energy performance
- EnergyPlus:
  Performs energy simulation, load calculation, building performance, simulation, energy performance, heat balance, and mass balance
- ENVSTD and LTGSTD:
  These programs calculate compliance with the federal commercial building standard, code compliance, and energy savings
- ERATES:
  Listing of electricity costs and electric utility rate schedules
- FRESA:
  A tool for assessing renewable energy and retrofit opportunities
- GS2000:
  A program to assess the benefits of geothermal heat pumps, heat exchanger sizing, and ground source heat pumps
- HEED:
  Performs whole building simulation, energy-efficient design, climate-responsive design, assesses energy costs and indoor air temperature
- HiLight:
  A program for assessing lighting and energy compliance
- HVACSIM+:
  Assists in assessing HVAC equipment, systems, controls, EMCS, and complex systems
- I-BEAM:
  Program involving indoor air quality, IAQ education, IAQ management, and energy consumption
- LISA:
  Performs life-cycle analysis, sustainability reviews, utilization, and embodied energy calculations

- Louver Shading:
  A program for assessing window, overhang, blinds, louvers, trellis, and shading
- MotorMaster+:
  Motors, energy-efficient motors, motor database, motor management, industrial efficiency
- Panel Shading:
  Assesses solar panels, photovoltaics, solar collectors, solar thermal, and shadings
- Photovoltaics Economics Calculator:
  Calculates the economic benefit of solar photovoltaic assemblies
- Quick Est:
  Performs lighting, 3D drawing, indoor lighting calculations
- Radiance:
  A rendering and calculation tool for assessing lighting and daylighting benefits
- RESEM:
  An analysis tool for retrofit work and institutional buildings
- RETScreen:
  Calculator for renewable energy screening and feasibility studies for energy efficiency
- SkyVision:
  Performs analysis of skylights, light well, fenestration, glazing, optical characteristics, and daylighting
- SPOT:
  Calculator for daylighting, electric lighting, photosensor, and related energy savings
- Star Performer:
  Tool for energy performance diagnostics and energy audits, mostly associated with office buildings
- Tariff Analysis Project:
  A utility rate tool that includes a bill calculator, utility bills review, tariff, schedules, rates, rate schedules, utility rates, utility tariffs, cost savings, energy-savings analysis, and investment analysis
- TOP Energy:
  A program that performs energy-efficiency optimization, simulation, variant comparison, and visualization of energy flows
- Visual:
  Program that assesses general lighting, lighting design, and roadway lighting using the lumen method
- WATERGY:
  A tool for assessing water conservation opportunities and potential energy savings

OTHER NONGOVERMENTAL WEB-BASED TOOLS

- Energy Depot: http://www.enercomusa.com/products.asp (Enercom)
- Energy Guide: http://www.energyguide.com (Nexus)
- EZ3W Weatherization Tool: http://ez3w.homeunix.com/ez3waudit/layout/frameset2.jsp

EZ3W is designed for weatherization professionals trained to evaluate existing homes using blower door tests and other measuring tools for assessing the building envelope.

■ Fuel Surcharge Index: http://www.fuelsurchargeindex.org/
Accurate fuel surcharge rates by run lane, run radius, and Interstate run based on fuel prices where a freight load will run.

■ ORNL Calculators: http://www.ornl.gov/sci/roofs+walls/calculators/index.html (Oak Ridge National Lab)

■ The Tariff Analysis Project (T.A.P.): http://tariffs.lbl.gov/
A comprehensive collection of electric utility tariff tools.

Government and Nonprofit Programs

Major state and federal government initiatives (see utility program listings later in this chapter)

■ U.S. State energy offices: www.naseo.org

■ U.S. State energy research organizations: www.asertti.org

■ U.S. State economic development agencies: www.nga.org

■ The California Energy Commission's Public Programs Office: http://www.energy.ca.gov/efficiency/public_programs.html
The California Energy Commission's Public Programs Office develops programs that promote energy efficiency in the local government, schools, hospitals, agricultural, industrial, and water treatment sectors of California.

■ Clean Cities Program: http://www.eere.energy.gov/cleancities/
The mission of the U.S. Department of Energy's Clean Cities Program is to advance the nation's economic, environmental, and energy security by supporting local decisions to adopt practices that contribute to the reduction of petroleum consumption. Clean Cities carries out this mission through a network of more than 80 volunteer coalitions, which develop public/private partnerships to promote alternative fuels and vehicles, fuel blends, fuel economy, hybrid vehicles, and idle reduction.

■ Florida Building Energy-Efficiency Rating System: http://www.fsec.ucf.edu/bldg/fyh/ratings/
In 1993, the Florida legislature enacted the Florida Building Energy-Efficiency Rating System Act. The first of its kind in the nation, this Act required the Department of Community Affairs to develop, adopt, and implement uniform, statewide energy rating systems for virtually all types of Florida buildings. The site provides information about the rating system for residential and commercial buildings as well as contact information.

■ New York Energy Smart Program: http://www.getenergysmart.org/
Energy-efficiency resources for New York state residents.

■ Rebuild America: http://www.eere.energy.gov/buildings/program_areas/rebuild.html

Rebuild America is a multiyear U.S. Department of Energy (DOE) program that helps community and regional partnerships improve commercial and multifamily building energy efficiency. Partnerships can be led by anyone. DOE provides partnerships with technical and financial assistance to help them carry out energy-efficiency retrofits.

■ Texas LoanSTAR Program: http://www.seco.cpa.state.tx.us/ls.htm
The LoanSTAR program is Texas' own program designed to "Save Taxes And Resources" by monitoring energy use and recommending energy-saving retrofits. In 1988, the Texas Governor's Energy Office (now known as the State Energy Conservation Office) received approval from the DOE to establish a statewide retrofit demonstration program. The initial capital came from oil overcharge funds. The LoanSTAR program is designed to demonstrate commercially available, energy-efficient, retrofit technologies and techniques. Part of the approved DOE program includes monitoring the buildings to determine the effectiveness of the conservation retrofits. The LoanSTAR Consortium is responsible for metering the buildings and analyzing the energy savings. LoanSTAR has already generated $15 million in savings (as of January 1995) for Texas taxpayers, and the program is projected to save another $250 million over the next 20 years.

■ U.S. DOE Federal Energy Management Program (FEMP): http://www1.eere.energy.gov/femp/
The federal government is the largest energy consumer in the nation. It used 1.92 quadrillion Btu (quads) of energy in fiscal year (FY) 1991, equal to 2.4 percent of all primary energy consumption in the United States at a cost of $11.28 billion. Federal energy consumption is widely dispersed, covering more than 500,000 buildings at 8000 locations worldwide. FEMP provides direction, guidance, and assistance to federal agencies in planning and implementing energy management programs that will improve the energy efficiency and fuel flexibility of the federal infrastructure.

■ U.S. Environmental Protection Agency (EPA)/DOE ENERGY STAR® Program: http://www.energystar.gov
ENERGY STAR® is a voluntary partnership among the U.S. EPA, U.S. DOE, product manufacturers, local utilities, home builders, retailers, and businesses. ENERGY STAR® encourages energy efficiency in products, appliances, homes, offices, and other buildings. Partners help promote energy efficiency by labeling products and buildings with the ENERGY STAR® logo, and educating consumers about the unique benefits of energy efficiency. In addition to promoting efficiency, ENERGY STAR® also offers tools to decrease operating costs, reduce air pollution, and save money for large and small businesses and organizations.

NONPROFIT ORGANIZATIONS

■ Leadership in Energy and Environmental Design (LEED): http://www.usgbc.org/
The LEED Green Building Rating System is a voluntary, consensus-based national standard for developing high-performance, sustainable buildings. Members of the U.S. Green Building Council representing all segments of the building industry developed LEED and continue to contribute to its evolution.

■ The Tax Incentives Assistance Project (TIAP): http://www.energytaxincentives.org/
The TIAP, sponsored by a coalition of public interest nonprofit groups, government
agencies, and other organizations in the energy-efficiency field, is designed to give
consumers and businesses information they need to make use of the federal income
tax incentives for energy-efficient products and technologies passed by Congress as
part of the Energy Policy Act of 2005 and subsequently amended several times.

INDUSTRY AND PROFESSIONAL ORGANIZATIONS

■ Air Conditioning Contractors of America (ACCA): http://www.acca-ncc.org/
ACCA is a nonprofit trade association of service to the independent heating, ventilation, air conditioning, and refrigeration (HVACR) contractors. Check out their Points
of Interest and Consumer Information page for help with selecting AC contractors,
how to make your home more comfortable, indoor air quality, and much more.
■ American Boiler Manufacturers Association (ABMA): http://www.abma.com/
ABMA is the national, nonprofit trade association of manufacturers of commercial/
institutional, industrial, and power-generating boilers; related fuel-burning equipment
and environmental systems; users of boiler and boiler-related equipment; and companies that provide products and services to the combustion equipment industry.
■ American Gas Association (AGA): http://www.aga.org/
AGA represents 195 local energy utility companies that deliver natural gas to more
than 56 million homes, businesses, and industries throughout the United States.
■ American Institute of Architects: http://www.aia.org
The American Institute of Architects is the national professional society of architects, active
in working with the International Code Council and U.S. Green Building Council.
■ American Petroleum Institute (API): http://api-ec.api.org/frontpage.cfm
API is the trade association of the oil and natural gas industry.
■ American Public Power Association (APPA): http://www.appanet.org/
APPA is the service organization for the nation's more than 2000 community-owned
electric utilities that serve more than 43 million Americans.
■ American Society of Heating, Refrigerating and Air-Conditioning Engineers
(ASHRAE): http://www.ashrae.org/
ASHRAE is an international organization of 50,000 persons with chapters throughout
the world. The Society is organized for the sole purpose of advancing the science
of HVACR for the public's benefit through research, standards writing, continuing
education, and publications.
   Through its membership, ASHRAE writes standards that create uniform methods
of testing and rating of HVACR equipment and establish accepted practices for the
industry worldwide. The Society's research program investigates numerous issues
related to industry practices, such as identifying new refrigerants that are environmentally safe. ASHRAE organizes broad-based technical programs for presentation at its
semiannual meetings and cosponsors the International Air-Conditioning, Heating,
Refrigerating Exposition, the largest HVACR trade show in North America.
   ASHRAE Bookstore: http://resourcecenter.ashrae.org/store/ashrae/

- American Solar Energy Society (ASES): http://www.ases.org/
  ASES is a national organization dedicated to advancing the use of solar energy for the benefit of U.S. citizens and the global environment.
- American Wind Energy Association (AWEA): http://www.awea.org/
  AWEA is a national trade association that represents wind power plant developers, wind turbine manufacturers, utilities, consultants, insurers, financiers, researchers, and others involved in the wind industry.
- Association of Energy Engineers (AEE): http://www.aeecenter.org/
  AEE is a society of over 8000 professionals involved in all areas of the energy field. AEE seeks to provide practical direction and reasonable solutions to the industry in the areas of energy management strategies, contingency planning, and alternatives for dealing with fuel price variations.
- Association of Energy Services Professionals (AESP): http://www.aesp.org/
  The Association of Energy Services Professionals (formerly the Association of Demand-Side Management Professionals) is an association of nearly 2000 utility executives, consultants, manufacturers, researchers, government regulators, and others concerned with promoting energy efficiency. All of the organization's members are active in the energy services industry in a wide variety of fields. Founded in 1989, the AESP conducts training courses, conferences, and workshops.
  AESP Publications List: http://www.fairmontpress.com/store/cbrandindex.cfm?id=3
- Chartered Institution of Building Services Engineers (CIBSE): http://www.cibse.org/
  CIBSE is the British counterpart to the ASHRAE. The files on the organization's website include links to UK sources and listings of their own publications, many of which are concerned with building energy efficiency.
- Cooling Technology Institute (CTI): http://www.cti.org/
  CTI's mission is to advocate and promote the use of environmentally responsible Evaporative Heat Transfer Systems (EHTS), cooling towers, and cooling technology for the benefit of the public by encouraging education, research, standards development and verification, government relations, and technical information exchange.
- Edison Electric Institute (EEI): http://www.eei.org/
  EEI is the premier trade association for U.S. shareholder-owned electric companies, serving international affiliates and industry associates worldwide. Their U.S. members service almost 95 percent of the customers in the shareholder-owned segment of the electric industry, as well as nearly 70 percent of all electric utility customers nationwide. The organization's members also generate over 70 percent of the electricity produced in the United States Organized in 1933, EEI represents its members' interests through advocating favorable policies in the legislative and regulatory arenas.
- El Paso Solar Energy Association (EPSEA): http://www. www.epsea.org
  EPSEA was founded in 1978 and is the oldest, continuously active, local solar organization in the United States. The purpose of EPSEA is to further the development and application of solar energy and related technologies with concern for the ecologic,

social, and economic fabric of the region (West Texas, Southern New Mexico, Northern New Mexico).

- Energy Efficient Building Association, Inc. (EEBA): http://www.eeba.org
The EEBA website offers online documents describing their mission, history, functions, programs, and membership options. The website also includes their criteria for designing, constructing, and rehabilitating residential and small commercial buildings.

- Energy Management Association of New Zealand: http://www.ema.org.nz
The Energy Management Association of New Zealand is dedicated to the promotion of efficient and sustainable energy use.

- Energy User News: http://www.energyusernews.com/
Energy User News, a Chilton business news magazine, reports on information for professional facility managers and their engineering staffs, including the consultants, contractors, and architects who work with them. Coverage includes all areas of managing energy-intensive building systems, including fuel and power acquisition.

- The Envirosense Consortium, Inc.: http://www.envirosense.org/
Envirosense Consortium, Inc. is a nonprofit organization of strategic companies and other associations that address Indoor Air Quality (IAQ) issues. The Consortium provides educational programs that focus on solutions for building, product, and maintenance systems.

- Global Network for Environmental Technology (GNET): http://www.gnet.org/
GNET contains information resources on environmental news, innovative technologies, government programs, contracting opportunities, market assessments and information, current events, and other material of interest to the environmental technology community.

- The Heat Pump Association (HPA): http://www.feta.co.uk/hpa/
HPA is the United Kingdom's leading authority on the use and benefits of heat pump technology and includes many of the country's leading manufacturers of heat pumps, components, and associated equipment.

- IEA Heat Pump Centre: http://www.HeatPumpCentre.org/
The IEA Heat Pump Centre (HPC) is the International Energy Agency's information center for heat pumping technologies, applications, and markets. The HPC operates worldwide via a network of national contacts in its member countries.

- Illuminating Engineering Society of North America (IESNA): http://www.iesna.org/
IESNA is home to over 70 technical, research, design, application, and educational committees that study and report on all aspects of lighting. IESNA committees set consensus standards and recommended practices that affect the public and lighting professionals. The organization holds LIGHTFAIR, their annual conference to learn about the latest activities and advances in the industry, and is the largest lighting trade show in the United States. The IESNA has over 90 local sections and 18 active student chapters throughout the United States, Canada, and Mexico.

- Institute of Electrical and Electronics Engineers (IEEE): http://www.ieee.org/portal/site
IEEE is the world's largest technical professional society. A nonprofit organization, they promote the development and application of electrotechnology and allied sciences for the advancement of the profession.

- International Code Council (ICC): http://www.iccsafe.org
  ICC publishes a comprehensive, coordinated set of building safety and fire prevention codes. ICC codes have been adopted in all 50 U.S. states, as well as several departments of the U.S. government and a number of foreign countries. The ICC is currently developing the International Green Construction Code (IgCC) to be published in early 2012. This code is expected to be widely adopted by jurisdictions as the first regulatory and adoptable green code.
- ICC-Evaluation Service (ICC-ES): www.icc-es.org
  ICC-ES is the U.S. leader in evaluating building products for compliance with code. A nonprofit, limited liability company, ICC-ES does technical evaluations of building products, components, methods, and materials. The evaluation process culminates with the issuance of reports on code compliance, which are made available free of charge to code officials, contractors, specifiers, architects, engineers, and anyone else with an interest in the building industry and construction. ICC-ES evaluation reports provide evidence that products and systems meet code requirements. A part of ICC-ES services is related to evaluation of products for their green/sustainability attributes, a program called SAVE (Sustainable Attributes Verification and Evaluation).
- International Commission on Illumination (CIE): http://www.cie.co.at/cie/
  The International Commission on Illumination—abbreviated as CIE from its French title Commission Internationale de l'Eclairage—is an organization devoted to international cooperation and exchange of information among its member countries on all issues relating to the science and art of lighting.
- International Electrotechnical Commission (IEC): http://www.iec.ch/
  The IEC was founded in 1906 for the purpose of promoting cooperation in the international standardization of electrical and electronic engineering. The IEC is composed of 49 national committees, representing all the industrial countries in the world.
- International Ground Source Heat Pump Association (IGSHPA): http://www.igshpa.okstate.edu/
  IGSHPA is a technical and marketing organization that promotes the greater use of geothermal heat pumps as an alternative to more commonly used products.
- International Standardization Organization (ISO): http://www.iso.org/iso/en/ISOOnline.frontpage
  ISO, established in 1947, is a nongovernmental worldwide federation of national standards bodies from approximately 100 countries. The mission of ISO is to promote the development of standardization with the goal of facilitating greater international exchange of goods and services as well as to enhance cooperation in the areas of intellectual, scientific, technological, and economic activity. ISO's work results in international agreements that are published as International Standards.
- National Association of Energy Service Companies (NAESCO): http://www.naesco.org/
  NAESCO's mission is to promote the delivery by energy service companies (ESCOs) of comprehensive energy services, including energy-efficiency environmental sustainability services. NAESCO disseminates information about developing technologies and their appropriate applications; participates in legislative and regulatory

proceedings that affect energy policy; and promotes best practices among ESCOs in the delivery of energy services.

■ National Association of Home Builders (NAHB): http://www.nahb.org/
NAHB Economic and Housing Data: http://www.nahb.org/page.aspx/category/sectionID=113
  NAHB produces in-depth economic analyses of the home building industry based on private and government data. Their Economics Group surveys builders, home buyers, and renters to gain insight into the issues and trends driving the industry. NAHB also hosts the Construction Forecast Conference, a twice yearly conference focusing on issues related to housing and the economy.

■ NAHB Home Builder Bookstore: http://store.builderbooks.com/cgi-bin/builder-books
NAHB's Home Builder Press and Bookstore were established to publish and sell professional development and educational products to the residential building industry.

■ NAHB's Research Center: http://www.nahb.org/page.aspx/generic/sectionID=96
The NAHB Research Center was founded in 1964 as a separately incorporated, not-for-profit subsidiary of NAHB. NAHB has 182,000 members, including more than 50,000 who build more than 80 percent of all U.S. homes. The Research Center keeps the U.S. home building industry informed of new technology and changing requirements.

■ NAHB's Journal—Builder Online: http://www.builderonline.com/
Builder Online is a professional resource to assist members in staying competitive within the homebuilding industry. The website contains databases, an online bookstore, home plans, and a listing of services and software.

■ National Association of Regulatory Utility Commissioners (NARUC): http://www.naruc.org/
NARUC promotes the advancement of commission regulation through the study and of subjects concerning the operation and supervision of public utilities and carriers. The organization also promotes: uniformity of regulation of public utilities and carriers, coordinated action by the commissions of the states with respect to the regulation of public utilities and carriers, and cooperation of state commissions with each other and with the federal commissions represented in the association.

■ National Petroleum Council (NPC): http://www.npc.org/
Formed in 1946, the NPC prepares reports that deal with virtually every aspect of oil and gas operations, including examinations of the ongoing and future operations and requirements of the U.S. oil and gas industries, statistical studies descriptive of these industries, delineations of the U.S. oil and gas resource base, and comprehensive analyses of the domestic energy supply/demand situation. Other studies have focused on environmental and energy conservation, technology, and legal issues.

■ National Propane Gas Association (NPGA): http://www.npga.org/i4a/pages/index.cfm?pageid=1
NPGA is the national trade association representing the U.S. propane industry. Their membership includes small businesses and large corporations engaged in the retail marketing of propane gas and appliances; producers and wholesalers of propane

equipment; manufacturers and distributors of propane gas appliances and equipment; fabricators of propane cylinders and tanks; and propane transporters.

■ National Roofing Contractors Association (NRCA): http://www.nrca.net/
NRCA, established in 1886, is one of the construction industry's oldest trade associations. NRCA mostly represents roof deck contractors and waterproofing companies. Its industry-related associate members include manufacturers, distributors, architects, consultants, engineers, and city and state/federal government agencies. NRCA has more than 4000 members and is affiliated with 99 state, regional, and international roofing contractor associations.

■ National Rural Electric Cooperative Association (NRECA): http://www.nreca .coop/
Founded in 1942, NRECA is the national service organization dedicated to representing the national interests of cooperative electric utilities and consumers in the United States. NRECA's more than 900 member cooperatives serve 40 million people in 47 states.

■ North American Electric Reliability Council (NERC): http://www.nerc.com/
NERC's mission is to promote the reliability of the electricity supply in North America. NERC helps electric utilities and other electricity suppliers cooperate by reviewing the past practices; monitoring members for compliance with current policies, criteria, standards; and assessing the future reliability of the bulk electric systems.

■ North American Insulation Manufacturers Association (NAIMA): http://www.naima .org/main.html
NAIMA is a trade association of North American manufacturers of fiber glass, rock wool, and slag wool insulation products. NAIMA members manufacture the vast majority of such products used in North America. Their website offers information on products used for commercial and industrial buildings.

■ Solar Energy Industries Association (SEIA): http://www.seia.org/
SEIA is the national trade association representing manufacturers, suppliers, and engineers in the photovoltaic and solar thermal industries. SEIA promotes the use and acceptance of solar technology in the world marketplace through publications, statistics, marketing data, policy development, education, and outreach to the general consumer. The organization holds SOLTECH, an annual solar conference.

■ Sustainable Buildings Industries Council (SBIC): http://www.sbicouncil.org/
SBIC is an independent, nonprofit organization whose mission is to advance the design, affordability, energy performance, and environmental soundness of U.S. buildings.

Utility Programs

GENERAL

■ U.S. Department of Energy (Federal Energy Management Program): http://www1 .eere.energy.gov/femp/financing/energyincentiveprograms.html
National database of electric load management and gas consumption management programs.

■ Lawrence Berkeley Laboratory (Environmental Energy Technologies Division): http://eetd.lbl.gov/EnergyCrossroads/2ueeprogram.html
List of public utility programs specifically oriented to small business/commercial usage.

PUBLIC UTILITIES

Commercial/business energy-efficiency programs are offered by public utilities around the United States. Some residential programs are listed as their incentives may be applicable to small businesses as well, particularly related to energy-efficient hot water heater and furnace installations.

Alabama

Alabama Power Company Information for Businesses: http://www.alabamapower.com

Tennessee Valley Authority (TVA) Information for Businesses: http://www.tva.gov/commercial/index.htm

City of Florence, Energy Right Program: http://www.florenceutilities.com/Utilities_Departments/Electricity/Energy_Right/index.html
Residential: Incentives for heat pump retrofit, and new electric water heaters.

Huntsville Utilities: http://www.hsvutil.org/news/conservation.shtml
Residential: Incentive programs for new energy-efficient HVAC systems.

Scottsboro Electric Power Board: http://www.scottsboropower.com/power/power5.html
Residential: TVA heat pump program and a water heater program.

Alaska

Anchorage Light and Power: http://www.mlandp.com/new%20paint/Safety_&_Efficiency.htm

Alaska Electric Light and Power: http://www.aelp.com/

Alaska Power Association (APA): http://www.areca.org/

Alaska Village Electric Cooperative, Inc. (AVEC): http://www.avec.org/index.html

Chugach Electric: http://www.chugachelectric.com/

Golden Valley Electric Association (GVEA): http://www.gvea.com/

GVEA's Business Sense web page: http://www.gvea.com/memserv/energysense/businesssense.php

Matanuska Electric Association: http://www.matanuska.com/

Renewable Energy Alaska Project (REAP): http://www.alaskarenewableenergy.org/

Arizona

Salt River Project: http://www.srpnet.com/menu/energy.aspx
Commercial: Business Energy Manager online survey, lighting efficiency analysis, information on energy-efficient equipment.

Arizona Power Services (APS) Information for Businesses: http://www.aps.com/ aps_services/business/waystosave/default.html

Southwest Gas Information for Businesses: http://www.swgas.com/commind/indust/ index.php?val=S

Arkansas

City of North Little Rock: http://www.nlrelectric.com/energy_index.htm
Residential/Commercial: Website contains energy-saving tips for both heating and cooling and appliances and lighting. They also have a downloadable "Energy Savers Guide" available.

Paragould City Light, Water & Cable Commercial: http://www.clwc.com/energy _efficiency.php
Website offers information on energy efficiency and links to ENERGY STAR product information.

Entergy Corporation Information for Businesses: http://www.entergyarkansas.com/ ar/yourbusiness/business_energy.asp

Empire District Electric Company: http://www.empiredistrict.com/

California

Alameda Power & Telecom Commercial: http://www.alamedapt.com/electricity/ com_savings.html
Lighting rebates, air conditioning rebates, energy-efficient building design and equipment grants, portable lending meter program, business energy audits.

Anaheim Public Utilities Commercial: http://www.anaheim.net/article.asp?id=990
Business energy incentives and rebates, energy-saving tips, green building information.

Azusa Light & Water Energy Conservation Tips: https://www.azusalw.com/Services/ Electric/conservation.asp, Service Options: https://www.azusalw.com/Services/ Customer/special.asp
Commercial: Financial incentive for the replacement or installation of central electric units with high-efficiency equipment to eligible customers; Commercial and Industrial Energy Partnership.

Bonneville Power Administration: http://www.bpa.gov/corporate/

Glendale Water and Power Commercial: http://www.ci.glendale.ca.us/gwp/Business _Services.asp

Smart Business energy audit and retrofit service, AC tune-up and duct testing, lighting rebates.

Imperial Irrigation District Commercial: http://www.iid.com/Energy_Index .php?pid=398
Rebates for efficient lighting and other efficient equipment.

Lassen MUD Residential: http://www.lmud.org/rebate.htm
Information about rebates offered by Lassen MUD to their customers for purchasing energy-efficient appliances, etc. Also has links to other energy-efficiency resources.

Lodi Electric Utility Commercial: http://www.lodielectric.com/commercial/
Energy rebates, energy-saving tips, and energy audits. Also has information about the PowerPrice Web application for commercial customers, which presents their energy data in an easy-to-read graphical format available for viewing, graphing, and analysis.

Long Beach Utility Information for Businesses: http://www.longbeach.gov/cd/econdev/default.asp

Los Angeles Department of Water and Power Commercial: http://www.ladwp.com/ladwp/cms/ladwp000580.jsp, Information for Businesses: http://www.ladwp.com/ladwp/cms/ladwp003154.jsp
Energy load monitoring program, lighting retrofits and rebates, HVAC rebates, customer generation rebate, premium efficiency motor program, chiller efficiency program, energy-efficiency financing, power quality consulting.

Merced Irrigation District Commercial: http://www.mercedid.org/energy/commercial/index.html
Information about energy efficiency and rebates and a few "success stories."

Modesto Irrigation District Commercial: http://www.mid.org/services/save/default .htm, Information for Businesses: http://www.mid.org/services/rebates/default.htm
Energy-efficient equipment rebates, energy audits.

Northern California Power Agency: http://www.ncpa.com/energy-efficiency-3.html
Commercial and Residential: Information on operational energy efficiency, demand-side energy efficiency, and cost-effective energy efficiency. Website also includes multiple reports and related documents on energy efficiency, including information about EPA's National Action Plan.

Pacific Gas and Electric Company (PG&E) Information for Businesses: http://www .pge.com/biz/rebates/

Pacific Power Information for Businesses: http://www.pacificpower.net/Homepage/Homepage35760.html

City of Palo Alto Commercial: http://www.cpau.com/bizindex.html
Rebates for installing PV systems, rebates for energy-efficient equipment, and Commercial Advantage Program [http://www.cpau.com/programs/ci-advantage/

cindex.html], which offers financial incentives to Palo Alto businesses to install state-of-the-art efficient equipment in place of aging and less energy-efficient equipment.

Pasadena Water and Power Commercial: http://www.ci.pasadena.ca.us/waterandpower/yourbusiness.asp

Rebates for energy-efficient equipment, incentives for design of green buildings, business newsletter, online energy analyzer, online energy information service

Riverside Public Utilities Commercial: http://www.riversideca.gov/utilities/business.asp

Energy-efficient equipment rebates, incentives for energy-efficient lighting, motors, and construction, custom energy-efficient technology grants, energy innovations grant program.

Roseville Electric Commercial: http://www.roseville.ca.us/electric/commercial_customers/default.asp, Information for Businesses: http://www.roseville.ca.us/electric/commercial_customers/commercial_rebates.asp

Business energy advisor service, rebates, energy-efficiency and energy-saving tips, Photovoltaic buy-down program, and shade tree program.

Sacramento Municipal Utility District Commercial: http://www.smud.org/commercial/saving/index.html

Rebates and financing for energy-efficient equipment, load cycling of air conditioners, online energy profiler, business energy analysis.

San Diego Gas and Electric Information for Businesses: http://www.sdge.com/business/

Sierra-Pacific Power Information for Businesses: http://econdev.sierrapacific.com/sppc/

Silicon Valley Power Commercial: http://www.siliconvalleypower.com/bus/?doc=bussave

Energy-efficient equipment rebates, new construction incentives, energy audits, online energy usage profiler.

Southern California Edison (SCE) Information for Businesses: http://www.sce.com/RebatesandSavings/SmallBusiness/

Southern California Gas (SoCal Gas) Information for Businesses: http://www.socalgas.com/business/

Southwest Gas Corporation Information for Businesses: http://www.swgas.com/commind/index.php

Truckee-Donner Public Utility District Commercial: http://www.tdpud.org/index.php?cId=34

Rebates.

Turlock Irrigation District Commercial: http://www.tid.org/tidweb/Bus/Rebates/index.htm

Offers rebates for chillers, lighting, motors, and refrigeration. Will provide technical design assistance and education to support the design and construction of energy-efficient facilities and process systems for new facilities or when retrofitting existing dairies.

Vernon Municipal Light Department: http://www.vernongov.org/coc-p.php?name=programBusDevr

Colorado

Colorado Springs Utilities Commercial: http://www.csu.org/environment/conservation_bus/index.html, Information for Businesses: http://www.csu.org/business/development/
   Rebates for energy-efficient equipment, incentives for builders to build green buildings, business newsletter, online energy analyzer, online energy information service.

Excel Energy Information for Businesses: http://www.xcelenergy.com/XLWEB/CDA/0,3080,1-1-3_19900-281-4_623_1129-0,00.html

Fort Collins Utilities: http://fcgov.com/utilities/powertosave/index.php
   Commercial: Rebates and incentives, design assistance for energy-efficient new buildings, and a newsletter.

Longmont Power & Communications Commercial: http://www.ci.longmont.co.us/lpc/bus/index.htm
   Energy-efficiency incentives and rebates.

Loveland Water and Power: http://www.ci.loveland.co.us/wp/power/Conservation/main.htm
   Commercial and Industrial: Provides funding for energy-efficient lighting retrofits, rebates for upgrading to energy-efficient cooling equipment, provides funding for energy-efficiency projects at commercial and industrial facilities, Automatic Load Profile Service.

Nebraska Municipal Power Pool (NMPP) Information for Businesses: http://www.nmppenergy.org/utilmgmt.htm

Platte River Power Authority Commercial: http://www.prpa.org/productservices/eepoverview.htm
   Energy Efficiency Program that provides funding for energy-efficiency projects at commercial and industrial facilities, also rebates/incentives for lighting and air conditioning.

Connecticut

Bozrah Light & Power: http://www.grotonutilities.com/elec_conserv.asp
   Commercial: Lighting program (both new installation and retrofit), air conditioning heat pump program, high-efficiency motor replacement program, vending and door heater control program, distributed generation and on-site energy solutions, and demand response programs.

Connecticut Light & Power Company Information for Businesses: http://www.cl-p.com/online/business/solutions/indexsolutions.asp

Norwich Public Utilities: http://www.norwichpublicutilities.com/efficiency-main.html
  Commercial: Prescriptive program for lighting, rebates, zero percent (0%) financing for qualifying commercial and industrial customers.

Southern Connecticut Gas Information for Businesses: http://www.soconngas.com/New%20Pages/New%2024.html

United Illuminating Company Information for Businesses: http://www.uinet.com/your_business/index.asp

Delaware

Chesapeake Utilities: http://www.chpkgas.com/

Delmarva Power: http://www.delmarva.com/dp/index.cfm

District of Columbia

Pepco: http://www.pepco.com/home/

Washington Gas: http://www.washgas.com/

Florida

Kissimmee Utility Authority: http://www.kua.com/Services/

Gainesville Regional Utility: http://www.gru.com/YourHome/Conservation/Energy/energysavers.jsp
  Commercial: Offers business rebates for solar electric, PV, and LED exit signs. Also provides services for smart vendor equipment and customized rebates for business energy-efficiency upgrades. http://www.gru.com/YourBusiness/Conservation/Energy/Rebates/lgBusiRebates.jsp

Florida Power & Light: http://www.fpl.com/

Gulf Power Company: http://www.southerncompany.com/gulfpower

Ocala Electric Utility: http://www.ocalaelectric.com/default.asp

Orlando Utilities Commission: http://www.ouc.com/

Progress Energy Florida, Inc.: http://www.progress-energy.com

TECO Peoples Gas: http://www.peoplesgas.com/

TECO Tampa Electric: http://www.tampaelectric.com/

Georgia

Georgia Power: http://www.southerncompany.com/gapower/home.asp

South Carolina Electric & Gas Corporation (SCE&G): http://www.sceg.com/en/

Tennessee Valley Authority (TVA): http://www.tva.gov/products/index.htm

Hawaii

Hawaiian Electric Company, Inc. (HECO): http://www.heco.com/portal/site/heco

Idaho

Bonneville Power Administration: http://www.bpa.gov/corporate/

Idaho Power Company: http://www.idahopower.com/

Illinois

Ameren: http://www.ameren.com/

Commonwealth Edison (ComEd): http://www.exeloncorp.com/ourcompanies/comed/

Indiana

Duke Energy: http://www.duke-energy.com/

Vectren: https://www.vectrenenergy.com/web/eenablement/frameset.jsp

Iowa

Iowa Association of Municipal Utilities: http://www.iamu.org/services/electric/efficiency/default.htm
   Produces Iowa Energizer Bulletin and other marketing materials on energy efficiency and renewable energy, offers commercial/industrial and residential auditor screening and take-home energy audits, participates in Marathon water heater program, and encourages use of biodiesel.

Alliant Energy: http://www.alliantenergy.com/docs/groups/public/documents/pub/default.hcsp, Information for Businesses: http://www.midwestsites.com/stellent2/groups/public/documents/pub/mws_os_000895.hcsp

Nebraska Municipal Power Pool (NMPP): http://www.nmppenergy.org/nmpp.htm

Kansas

Empire District Electric Company: http://www.empiredistrict.com/

Excel Energy: http://www.xcelenergy.com/XLWEB/CDA/0,3080,1-1-1_18554-127-0_0_0-0,00.html

Kansas City Power & Light (KCP&L): http://www.kcpl.com/

Kansas Electric Power Cooperative, Inc. (KEPCO): http://www.kepco.org/alternate/index.html

Nebraska Municipal Power Pool (NMPP): http://www.nmppenergy.org/nmpp.htm

Kentucky

Duke Energy: http://www.duke-energy.com/

KU Energy Corporation: http://www.lgeenergy.com/rsc/ku/default.asp

LG&E Energy: http://www.lgeenergy.com/lge/default.asp

Tennessee Valley Authority (TVA) Information for Businesses: http://www.tva.gov/products/index.htm

Louisiana

Entergy Corporation: http://www.entergy-louisiana.com/la/default.asp

Southwestern Electric Power Company: http://www.swepco.com/

Maine

Central Maine Power: http://www.cmpco.com/

Maryland

Baltimore Gas & Electric: http://www.bge.com/portal/site/bge

Delmarva Power: http://www.delmarva.com/dp/index.cfm

Pepco Energy Services: http://www.pepco-services.com/

Washington Gas: http://www.washgas.com/

Massachusetts

Braintree Electric Light Department: http://www.beld.com/Electric/Programs.asp
Rebates for energy-efficient applications, appliance cost calculator, alternative energy projects, conservation education, Energy Advisor/technical assistance, load management.

Taunton Municipal Lighting Plant Commercial: http://www.tmlp.com/commercial.html

KeySpan Energy: http://www.keyspanenergy.com/corpinfo/about/index_all.jsp

Massachusetts Electric: http://www.nationalgridus.com/masselectric/about_us/aboutus.asp

NSTAR: http://www.nstaronline.com/business/

Michigan

CMS Energy: http://www.cmsenergy.com/mst/

Consumer's Energy Company: http://www.consumersenergy.com/welcome.htm

DTE Energy: http://www.dteenergy.com/

Excel Energy: http://www.xcelenergy.com/

Lansing Board of Water and Light, Michigan: http://www.lbwl.com/

Northern Indiana Public Service Company: http://www.nipsco.com/

Wisconsin Public Service Corporation: http://www.wpsc.wpsr.com/

Minnesota

Marshall Municipal Utilities Commercial and Industrial: http://www.marshallutilities
.com/business/rebates.php
   Rebates for replacing inefficient lighting with new high-efficiency lighting, high-
efficiency air conditioning, and heat pump rebates. Infrared imaging can also be used
to locate energy losses in buildings.

Excel Energy: http://www.xcelenergy.com/

Minnesota Power: http://www.mnpower.com/about_mp/index.htm

Mississippi

Entergy Corporation: http://www.entergy-mississippi.com/ms/default.asp

Mississippi Power Company: http://www.southerncompany.com/mspower/home
.asp

Tennessee Valley Authority (TVA) Information for Businesses: http://www.tva.gov/
products/index.htm

Missouri

Ameren: http://www.ameren.com/

Acquila, Inc.: http://www.aquila.com/

Empire District Electric Company: http://www.empiredistrict.com/

Kansas City Power & Light: http://www.kcpl.com/

City Utilities of Springfield: http://www.cityutilities.net/conserve/overview
.htm
   Commercial: programmable thermostat rebate, irrigation system rain sensor rebate,
online business energy audits, online business energy library, on-site commercial
energy audits, on-site commercial lighting analysis, commercial lighting rebate for
high bay, commercial lighting rebate for T12 to T8 conversions.

Columbia Water & Light Commercial: http://www.gocolumbiamo.com/
WaterandLight/
   Engineering, lighting, and power analysis, distributed generation, low-interest
loans for efficiency improvement, rate savings and load shedding programs.

Independence Power & Light Residential: http://www.ci.independence.mo.us/pl/
Energy-saving tips, rebates for energy-efficient appliances.

Montana

Bonneville Power Administration: http://www.bpa.gov/Energy/N/

Energy West: http://www.ewst.com/montana.html

Northwestern Energy: http://www.northwesternenergy.com/

Nebraska

Nebraska Public Power District Commercial: http://www.nppd.com/My_Business/
Commercial_Services/Additional_Files/energy_solutions.asp
Energy audits, indoor and outdoor lighting assessments, motor efficiency optimization, infrared thermal imaging, energy demand and rate optimization, heating and cooling system comparison.

Nebraska Municipal Power Pool (NMPP) Information for Businesses: http://www
.nmppenergy.org/utilmgmt.htm

Northwestern Energy: http://www.northwesternenergy.com/

Omaha Public Power District: <http://www.oppd.com/

Nevada

Bonneville Power Administration: http://www.bpa.gov/corporate/About_BPA/

Sierra Pacific: http://www.sierrapacific.com/

Southwest Gas: http://www.swgas.com/

New Hampshire

Granite State Electric: http://www.nationalgridus.com/granitestate/about_us/aboutus.asp

Public Service of New Hampshire: http://www.psnh.com/AboutPSNH/default.asp

New Jersey

Atlantic City Electric: http://www.atlanticcityelectric.com/ace/index.cfm

Public Service Electric & Gas (PSE&G): http://www.pseg.com/about/company
_overview.jsp

New Mexico

Excel Energy: http://www.xcelenergy.com/

PNM: http://www.pnm.com/about/home.htm

Texas-New Mexico Power Company: http://www.tnpe.com/aboutus.asp

New York

Central Hudson Gas & Electric Corporation: http://www2.centralhudson.com/

KeySpan Energy: http://delivery.keyspanenergy.com/index.jsp

Long Island Power Authority: http://www.lipower.org/company/

Niagara Mohawk: http://www.nationalgridus.com/niagaramohawk/

Rochester Gas & Electric Corporation (RGE): http://www.rge.com/rgeweb/webcontent .nsf/doc/homepage

North Carolina

Elizabeth City Electric Department
   Offers free energy audits, "Cycle and Save" load management program, online home energy conservation tools through Energy Depot.

Duke Energy: http://www.duke-energy.com/

Lockhart Power Company: http://www.lockhartpower.com/

Progress Energy: http://www.progress-energy.com/aboutus/index.asp

South Carolina Electric & Gas Corporation (SCE&G): http://www.sceg.com/en/

Tennessee Valley Authority (TVA) Information for Businesses: http://www.tva.gov/ products/index.htm

North Dakota

Excel Energy: http://www.xcelenergy.com/

North Dakota Association of Rural Electric Cooperatives (NDAREC): http://www .ndarec.com/

Nebraska Municipal Power Pool (NMPP): http://www.nmppenergy.org/nmpp .htm

Otter Tail Power Company: http://www.otpco.com/default.asp

Ohio

Bryan Municipal Utilities: http://www.cityofbryan.net/Default.asp
   Distributes compact fluorescent light bulbs (CFLs) to customers, offers online energy profile, conducts city department energy audit, and implements conservation measures.

American Electric Power: http://www.aepohio.com/

Cinergy: http://www.cinergy.com/

Dayton Power & Light (DP&L): http://www.waytogo.com/

First Energy: http://www.firstenergycorp.com/index.html

Oklahoma

Empire District Electric Company: http://www.empiredistrict.com/

Excel Energy: http://www.xcelenergy.com/

OGE Energy Corporation: http://www.oge.com/

Oklahoma Natural Gas (ONG): http://www.oneok.com/ong/ong_home.jsp

Public Service Company of Oklahoma (part of AEP): http://www.psoklahoma.com/about/default.asp

Oregon

Eugene Water and Electric Board Commercial: http://www.eweb.org/business/energy/index.htm
    Technical assistance, loans, and financial incentives for energy-efficient equipment, electric motors, training and seminars.

Bonneville Power Administration: http://www.bpa.gov/corporate/

Eugene Water and Electric Board: http://www.eweb.org/index.htm

NW Natural: https://www.nwnatural.com/index.asp

Portland General Electric Company (PGE): http://www.portlandgeneral.com/Default.asp?bhcp=1

Pennsylvania

First Energy: http://www.firstenergycorp.com/index.html

Peco Energy (part of Exelon Energy Delivery): http://www.exeloncorp.com/ourcompanies/peco/

PPL Corporation: http://www.pplweb.com/about/

Rhode Island

Narragansett Electric (now part of National Grid): http://www.nationalgridus.com/narragansett/new_name.asp

South Carolina

Progress Energy: http://www.progress-energy.com/aboutus/index.asp

Duke Energy Corporation: http://www.duke-energy.com/

South Carolina Electric & Gas Corporation (a SCANA company): http://www.sceg.com/en/

South Dakota

Excel Energy: http://www.xcelenergy.com/

Northwestern Energy: http://www.northwesternenergy.com/

Otter Tail Power Company: http://www.otpco.com/default.asp

Tennessee

Tennessee Valley Authority (TVA) Information for Businesses: http://www.tva.gov/products/index.htm

Texas

Austin Energy Commercial: http://www.electric.austin.tx.us/Commercial/index.htm
   Offers rebates for efficient appliances, a green building program, and a load profiler where you can view your energy use.

Entergy Texas: http://www.entergy-texas.com/

Excel Energy: http://www.xcelenergy.com/

Green Mountain Energy: http://www.greenmountain.com/

Reliant Energy: http://www.reliant.com/reliant/

Texas Electric Choice: http://www.powertochoose.org/

Texas-New Mexico Power Company Information for Businesses: http://www.tnpeefficiency.com/

TXU Energy: http://www.txu.com/Cultures/en-US/default.htm

Utah

Bonneville Power Administration (BPA): http://www.bpa.gov/corporate/

Utah Power (a PacificCorp company): http://www.utahpower.net/Homepage/Homepage35888.html

Vermont

Central Vermont Public Service Corporation (CVPS): http://www.cvps.com/

Green Mountain Power: http://www.gmpvt.com/

Vermont Gas: http://www.vermontgas.com/

Virginia

Delmarva Power: http://www.delmarva.com/home/

Dominion Virginia Power: http://www.dom.com/about/companies/vapower/index.jsp

Tennessee Valley Authority (TVA) Information for Businesses: http://www.tva.gov/products/index.htm

Washington Gas Company: http://www.washingtongasliving.com/index.cfm

Washington

Bonneville Power Administration: http://www.bpa.gov/Energy/N/

NW Natural: https://www.nwnatural.com/index.asp

Pend Oreille Public Utility District: http://www.popud.com/energy.htm

Puget Sound Energy: http://www.pse.com/index.html

Seattle City Light Commercial: http://www.seattle.gov/light/conserve/business/
  Financial incentives for energy-efficient equipment and lighting and vending machines, technical assistance.

Snohomish PUD Commercial: http://www.snopud.com/Default.ashx?p=1805
  Incentives for energy-efficiency conservation strategies

Tacoma Power: http://www.ci.tacoma.wa.us/power/
  Commercial: Rebates for efficient commercial refrigerators, freezers, fluorescent lighting, and transformers, in addition to a resource center.

Washington's Public Utilities Districts: http://www.wpuda.org/pudlinks.htm

West Virginia

Allegheny Power: http://www.alleghenypower.com/

American Electric Power (AEP): http://www.aep.com/

Appalachian Power: http://www.appalachianpower.com/

Wisconsin

Wisconsin Public Power, Inc.: http://www.wppisys.org/
  Offers grants for energy-efficiency projects to commercial and industrial customers through the RFP for Energy Efficiency.

Excel Energy: http://www.xcelenergy.com/Company/Pages/Home.aspx

Focus on Energy: http://www.focusonenergy.com/

Madison Gas & Electric: http://www.mge.com/

Minnesota Power: http://www.mnpower.com/

Northern Indiana Public Service Company: http://www.nipsco.nisource.com/

WE Energies: http://www.we-energies.com/

Wisconsin Energy Corporation: http://www.wisconsinenergy.com/

Wisconsin Public Service Corporation: http://www.wpsc.wpsr.com/

Wyoming

Bonneville Power Administration: http://www.bpa.gov/

Cheyenne Light, Fuel and Power: http://www.cheyennelight.com/

Nebraska Municipal Power Pool (NMPP): http://www.nmppenergy.org/nmpp.html

Pacific Power: http://www.pacificpower.net/Homepage/Homepage35759.html

Wyoming Municipal Power Agency: http://www.wmpa.org/

GLOSSARY

Architectural salvage: Architectural salvage refers to existing building materials or features removed by a salvage contractor before or during the demolition process. Architectural salvage value is considered to be waste that is diverted from landfills.

Baler: Baler is a machine that compacts waste materials, usually into rectangular bales. Balers often are used on newspaper, plastics, and corrugated cardboard. Balers compress saleable ferrous scrap into a more uniform, rectangular shape (sometimes referred to as a "log") and enable scrap metal recyclers to move baled material more efficiently from the yard to the shredder or to the mill.

Btu (British thermal unit): Btu is the amount of heat energy needed to heat 1 pound of water at $1°F$ ($-17°C$).

Building permit: It refers to the approval issued by the appropriate local building or construction department allowing construction of the project to proceed. Building permits are often issued for various components of the work, including: demolition; foundation/footing; fire protection; electrical; and plumbing. It is also known as a construction permit.

By-product: By-product is a secondary product of a manufacturing process. A waste by-product is an unwanted by-product that can either be disposed of or recycled.

Ceiling tiles: These panels, also called acoustical ceiling tiles, are made from a variety of materials and are designed to reduce noise. Many recyclers will accept mineral fiber-based tiles that are free of contaminants but will not normally accept cast tile, fiberglass board, ceramic-based, or laminated tiles.

Clean rubble: Clean rubble is inert, uncontaminated construction and demolition waste, which can include asphalt, concrete and concrete products, reinforcing steel, brick and concrete masonry units, soil, or rock.

Commercial waste: It is waste material that originates in wholesale business establishments, office buildings, stores, schools, hospitals, and government agencies. It is also known as retail waste.

Commingled recycling: It is a type of recycling that allows contractors to put types of waste into common containers to save space and labor. The commingled materials, typically consisting of wood, metal, and cardboard, are separated at the recycling facility for processing. Commingled recycling is less expensive than landfill disposal but typically more expensive than source-separated recycling. This type of recycling also makes it more difficult to accurately track material recycling and disposal by material type.

Compost: It is decomposed organic material resulting from the composting process. It is used to enrich or improve the consistency of soil.

Computerized Maintenance Management System (CMMS): CMMS is a computer system that measures, manages, and analyzes the maintenance processes at a facility.

Construction and demolition waste (C&D waste): It refers to solid waste resulting from the construction, remodeling, repair, and demolition of structures, roads, sidewalks, and utilities. Such wastes include, but are not limited to: bricks, concrete, and other masonry materials; roofing materials; soil; rock; wood and/or wood products; wall or floor coverings; plaster; drywall; plumbing fixtures; electrical wiring; electrical components containing no hazardous materials; nonasbestos insulation; and construction-related packaging. C&D waste does not include waste material containing friable asbestos; garbage; furniture and appliances from which ozone-depleting chlorofluorocarbons have not been removed in accordance with the provisions of the federal Clean Air Act; electrical equipment containing hazardous materials; tires; drums; and containers, even though such wastes resulted from construction and demolition activities. Clean rubble that is mixed with other C&D waste during demolition or transportation is normally considered to be C&D waste.

Construction Specifications Institute (CSI) Divisions of Construction: CSI is a national association promoting consistency and professionalism in the writing of construction specifications. The organization publishes the CSI Divisions of Construction, which is widely used as the organizational format for master specifications and construction cost-estimating systems.

Container rental: It refers to the monthly fee for having a compactor or Dumpster on-site.

Corrugated cardboard [also known as old corrugated cardboard (OCC) and/ or cardboard box]: It is a paper product made of unbleached kraft fiber, with two heavy outer layers and a wavy inner layer to provide strength. It is commonly used as a shipping container and is easily recyclable.

Current asset value (CAV): CAV is the current cost required to reproduce a specific asset. CAV is often used as a metric for CMMS implementations.

Deconstruction (also referred to as "soft demolition"): Deconstruction is often considered a sustainable alternative to conventional building demolition. In deconstruction, hazardous materials are removed, reusable building materials are salvaged, demolition materials are recycled, and only a small portion of waste ends up in the landfill. Deconstruction is also a specific phase of demolition in which materials suitable for reuse or recycling are removed from the building prior to general demolition.

Demolition debris: Demolition debris is waste resulting from demolition operations on pavements, buildings, or other structures that includes lumber, drywall, concrete, pipe, brick, glass, electrical wire, and rubble.

Densification: Densification is the process of packing recyclables closely together, such as baling or rerolling, to facilitate shipping and processing.

Diversion rate: It is a measure of the amount of waste being diverted from the municipal solid waste stream, either through recycling or composting.

Drywall: Drywall is an internal wall material made of gypsum, often covered on both sides with a paper facing. Drywall is also referred to as gypsum board, wallboard, plasterboard, and Sheetrock™.

Eco-industrial park: It refers to a community of manufacturing and service businesses located together on a common property using each others' wastes as materials for production.

Electronic waste: Sometimes referred to as e-waste, this is a term applied to consumer and business electronic equipment that is no longer useful. E-waste includes computers, televisions, radios, CD players, and other electronic equipment.

Embodied energy: Embodied energy is the sum total of the energy necessary—from raw material extraction, transport, manufacturing, assembly, and installation, plus the capital, environmental, and other costs—used to produce a service or product from its beginning through its disassembly, deconstruction, and/or decomposition.

Enterprise asset management (EAM): EAM refers to monitoring and managing an organization's assets across departments, locations, facilities, or business units.

Equipment lifetime: It is the span of time over which a piece of equipment is expected to fulfill its intended purpose.

Facility Condition Index (FCI): The Facility Condition Index is a facility management tool that is used to assess the current and projected condition of a building asset. The FCI is often the output from a Computerized Maintenance Management System (CMMS). The FCI is defined as the ratio of the current year's required renewal cost to the current building replacement value.

Facility management: It is the interdisciplinary field responsible for the maintenance and care of commercial or institutional buildings, including the management of the following building systems or services: heating, ventilating, and air conditioning (HVAC); utilities; plumbing; electrical; cleaning; security; and landscaping.

Ferrous metal: Ferrous metal is metal containing iron (such as steel) in sufficient quantities to allow for magnetic separation.

Forest residues: Forest residues are fibrous by-products of harvesting, manufacturing, extractive, or woodcutting processes.

Gaylord: It is a 1.4-cubic-yard (1-cubic-meter) cardboard container used to store loose materials.

Generation data: It is the information on waste amounts derived from actual waste materials produced, usually determined by assessing waste bins on-site.

Granulator: Granulator is a machine that produces small plastic particles.

Green building: It is the standard for construction that minimizes the effect of the built environment on the natural and social landscape.

Green Globes™: Green Globes is an environmental assessment, education, and rating system that is promoted in the United States by the Green Building Initiative, a Portland, Oregon–based nonprofit organization.

Greenwashing (also known as faux green): Greenwashing is to falsely claim that a product is environmentally sustainable.

Handler: Handler is a company that performs at least one of the following processes on collected recyclables: sorting, baling, shredding, or granulating.

Hauler: Hauler is a company that provides transportation (hauling) services and vehicles from the construction site to a recycling center.

Heating, ventilating, and air-conditioning (HVAC): HVAC refers to the building systems and accessories required to provide thermal comfort to the building's occupants.

High-density polyethylene (HDPE): HDPE is a plastic resin commonly used to make milk jugs, detergent containers, and base cups for plastic soda bottles. The standard plastic code for HDPE is number 2.

High-grade waste paper: It is the most valuable waste paper for recycling. High-grade waste paper can be substituted for virgin wood pulp in making paper. Examples of high-grade waste paper include letterhead stationery and computer paper.

ICC-ES SAVE program: International Code Council (ICC) Sustainable Attributes Verification and Evaluation™ (SAVE™) program is a program created to verify manufacturers' claims regarding the sustainable attributes of their products. Product evaluation under this program results in a Verification of Attributes Report™, which provides technically accurate product information that can be helpful to those seeking to qualify for points under major green rating systems.

Following are inappropriate for processing: loads of C&D materials entering a facility of which less than 90 percent, as determined by the processor, can be sent by the facility for recycling. Such loads are usually charged at a higher rate.

Industrial scrap: Industrial scrap refers to recyclables generated by manufacturing processes, such as trimmings and other leftover materials, or recyclable products that have been used by industry but are no longer needed, such as buckets, shipping containers, signs, pallets, and wraps.

Inert debris: It refers to those materials that are virtually inert, such as rock, dirt, brick, concrete, or other rubble.

Insulation: Insulation comes in many different forms. Fiberglass is the most common. It consists of flexible fragments of spun or woven glass formed into batts, though it is sometimes also blown into cavities. Cellulose insulation is also very common and usually consists of shredded newspaper mixed with a binder and fire retardant. Another type of insulation, vermiculite insulation, is a pebble-like, pour-in product that is light brown or gold in color. Vermiculite insulation sometimes contains asbestos fibers.

International Code Council (ICC): ICC is a membership association dedicated to building safety, fire prevention, and energy efficiency, and develops the codes used to construct residential and commercial buildings, including homes and schools. Most U.S. cities, counties, and states that adopt codes utilize ICC codes. ICC is working to develop the International Green Construction Code (IgCC) and the ICC-ES SAVE program. Model codes published by the ICC regulate building construction for safety, energy conservation, and sustainability. The ICC family of I-codes includes the International Building Code (IBC), International Energy Conservation Code (IECC), International Green Construction Code (IgCC to be published in 2012), International Residential Code (IRC), International Fire Code (IFC), International Plumbing Code (IPC), International Mechanical Code (IMC), International Private Sewage Disposal Code (IPSDC), International Fuel Gas Code (IFGC), International Existing Building Code (IEBC), International Property Maintenance Code (IPMC), International Zoning Code (IZC), International Wildland-Urban Interface Code (IWUIC), and ICC Performance Code for Buildings and Facilities (ICCPC).

International Facilities Management Association (IFMA): IFMA is the largest and most widely known professional association for facility managers.

International Green Construction Code (IgCC): IgCC is a model green construction code developed by the ICC in collaboration with other national building-related organizations. The IgCC is coordinated with the existing family of International Codes that span the spectrum of the industry from building to energy conservation. (See also International Code Council.)

Kilowatt (kW): The flow rate of electrical energy measured in 1000 watt (1 kilowatt) units. kW is used to measure the demand component of electric bills. It can also be used to designate the electrical output of generation systems.

Kilowatt hour (kWh): Kilowatt hour designates the amount of electrical energy used by an appliance or produced by a generation system within 1 hour. It is the standard unit of energy used for electric bill calculations.

LEED™: The acronym for Leadership in Energy and Environmental Design, it is a green building rating criteria developed by the U.S. Green Building Council.

The LEED rating system is nationally recognized as the primary green building standard.

Life cycle: Life cycle refers to the stages of a product, beginning with the acquisition of raw materials, continuing with manufacture, construction, and use, and concluding with a variety of recovery, recycling, or waste management options.

Locally sourced materials: These are materials obtained from within a defined radius around a project site, in order to support the local economy and to reduce transportation costs and energy.

Metrics: It is the objective means of measuring performance and effectiveness. It is also known as a Key Performance Indicator (KPI).

Mixed C&D waste: It refers to C&D materials containing both recyclable and nonrecyclable C&D materials that have not been source separated. C&D waste is considered to be "mixed" C&D waste if it contains more than 10 percent—but less than 90 percent—recyclable C&D waste by volume. At a mixed C&D recycling facility, different recyclables are sorted from a load of mixed debris. A load of mixed C&D generally includes drywall, metal, untreated wood, yard trimmings, and small amounts of inert materials.

Mixed paper: Mixed paper refers to waste paper of various kinds and qualities. Examples include stationery, notepads, manila folders, and envelopes.

Net present value (NPV): NPV is the current value of future revenue based on the time value of money.

Nonferrous metals: These are metals such as copper, brass, bronze, aluminum bronze, lead, pewter, zinc, and other metals to which a magnet will not adhere.

Nonrenewable resources: These are natural materials that are considered finite because of their scarcity, the long time required for their formation, or their rapid depletion.

Operations and maintenance (O&M): O&M typically includes the day-to-day activities required for the building, including its systems and equipment, to perform its intended function.

Organic waste: Organic waste refers to discarded living material, such as vegetative and food waste.

Pallet: Pallet is a wooden platform placed underneath large items so they may be picked up and moved by a forklift.

Paperboard: Paperboards are heavyweight grades of paper commonly used for packaging products like cereal boxes. Paperboard is classified differently from corrugated cardboard for recycling purposes.

PCBs: Polychlorinated biphenyls are a class of industrial chemicals manufactured from 1930 to 1977 for use in electrical and hydraulic products. PCBs are still present in the environment because of their persistence and accumulation, and may be encountered in items removed from demolished buildings.

Postconsumer: The term is used to describe types of recycled-content products containing materials that have been previously used by consumers and then reprocessed into new products.

Postindustrial recycled content: The term indicates that manufacturing waste has been cycled back into the production process. These products do not represent the significant resource savings that postconsumer products do, but are far preferable to those that use virgin materials.

Postmanufacture content: Also known as postmanufacture waste, this term refers to waste that was created by a manufacturing process and is subsequently only used as a constituent in another manufacturing process. It is similar to postindustrial waste, although that term is typically applied to waste products recycled within the same process.

Preconsumer (also known as postindustrial recycled content): These products contain waste materials created as a result of manufacturing processes that were then used in the manufacture of new products.

Pure loads of recyclable C&D waste: These are loads of single-type or mixed types of recyclable C&D waste that contain at least 90 percent recyclable C&D waste materials by volume.

Reclaimer (reprocessor): Reclaimer is a company that performs at least one of the following processes on collected recyclables: washing/cleaning, pelletizing, or manufacturing a new product.

Recyclables: Recyclables are any materials that will be used or reused as an ingredient in an industrial process to make a product, or as an effective substitute for a commercial product. Common construction recyclables include paper, glass, concrete, plastic, steel, and asphalt.

Recycle: It refers to the separation of construction waste or demolition materials into separate recycle categories for reuse into marketable materials. Examples of recycling include separating wood waste for recycling into paper pulp, or taking soil to a topsoil facility for reprocessing into topsoil.

Recycled content: It refers to the amount of a product's or package's weight that is composed of materials that have been recovered from waste. Recycled content may include preconsumer and postconsumer materials.

Replacement asset value (RAV): RAV is the monetary value that would be required to replace the production capability of the present assets in the plant.

Resource Conservation and Recovery Act (RCRA): RCRA was enacted by Congress in 1976 as an amendment to the 1965 Solid Waste Disposal Act. The goals of RCRA are to protect human health and the environment from the hazards posed by waste disposal, conserve energy and natural resources through waste recycling and recovery, reduce the amount of waste generated, and ensure that wastes are managed in a manner that is protective of human health and the environment.

Return on assets (ROA): ROA is profit divided by asset value.

Return on investment (ROI): ROI is the profit gained from an investment divided by the monetary value of the investment.

Reuse: It refers to the use of demolished materials on the same site for the same or different purpose. Examples include grinding concrete or asphalt for reuse on-site, or reusing framing lumber and steel sections.

Salvage: It is the act of removing construction or demolition waste from an existing building to be reused on that site or elsewhere in the same form, or the products removed as a result of salvage. Examples of salvage include removing brick, tile, ornamental architectural items, lumber, doors, and plumbing fixtures.

SAVE: Sustainable Attributes Verification and Evaluation, a program of the International Code Council Evaluation Service (ICC-ES) for an independent verification of building product sustainable attributes.

Scrap: Scrap refers to discarded or rejected industrial waste material often suitable for recycling.

Segregation: It is the systematic separation of solid waste into designated categories for pickup by recyclers.

Solid Waste Disposal Act: It is a federal law passed in 1965 and amended in 1970 that addresses waste disposal methods, waste management, and resource recovery.

Solid waste processing facility: It is an incinerator, composting facility, household hazardous waste facility, waste-to-energy facility, transfer station, reclamation facility, or any other location where solid wastes are consolidated, temporarily stored, salvaged, or otherwise processed prior to being transported to a final disposal site. This does not include scrap metal material recycling and processing facilities.

Source reduction: It refers to minimizing waste at the source of generation, preventing waste before it is generated.

Source separation: Source separation involves segregating recycled material into separate containers for pickup by a recycler and transfer to a facility for processing. Source separation requires more site area and individual containers, but makes tracking and quality control of the materials easier. Source separation is normally used for materials such as drywall, concrete, carpet, plastics, and ceiling tiles.

Strategic facility plan: It is a two- to five-year facility plan, covering the entire portfolio of owned and/or leased space, that maps the strategic facility improvement goals in support of the organization's overall strategic goals.

Substitution: Substitution is a product or process proposed by the contractor in lieu of the specified product or process, offered either with a credit for a lesser product or with no credit for an equivalent product. Typically, a substitution must be expressly approved by the architect prior to implementation or installation.

Sun hours: It refers to the average number of hours per day of usable solar radiation. In 1 hour, under ideal conditions, 1 square meter receives the equivalent of approximately 1 kilowatt hour of solar energy.

Trash (garbage): Trash is material considered worthless, unnecessary, or offensive that is usually thrown away; any product or material unable to be reused, returned, recycled, or salvaged.

Universal wastes: These are wastes that are hazardous upon disposal but pose a lower risk to people and the environment than other hazardous wastes do. State and federal regulations identify which unwanted products are universal wastes, and provide simple rules for handling and recycling them. Examples of universal wastes are televisions, computers, computer monitors, batteries, and fluorescent lamps.

Vinyl (V): It is a common type of plastic used to make shampoo bottles and other containers. The standard plastic code is number 3.

Virgin resources: These are resources using raw materials that have not been used before.

Volatile organic compounds (VOCs): VOCs are a principal component in atmospheric reactions that form ozone and photochemical oxidants. VOCs are emitted from diverse sources, including automobiles, chemical manufacturing facilities, dry cleaners, paint shops, and other commercial and residential sources that use solvent and paint.

Waste: It refers to material that has been discarded because it has worn out, is used up, or is no longer needed.

Waste audit: Waste audit is an inventory of the amount and type of solid waste that is produced at a specific site as a result of construction or demolition activities.

Waste exchange: It refers to two or more companies exchanging materials that would otherwise be discarded. The term also refers to an organization with electronic and/or catalog networks to match companies that want to exchange their materials.

Waste reduction: It is a construction site management strategy that encourages workers and subcontractors to generate less trash through practices such as reuse, recycling, and buying products with less packaging.

Waste reduction and recycling plan: It is a written plan for the recycling of project C&D debris.

Wetlands: The term refers to a lowland area, such as a marsh or swamp, saturated with water. Wetlands are crucial wildlife habitats, and important for flood control and maintaining the health of surrounding ecosystems.

INDEX